What mathematicians and teachers write about *FROM ZERO TO INFINITY*

"I read *From Zero to Infinity* when I was a schoolboy in Oxford, England, and my only regret is that I was well into my teens (17) before it happened. Just last week I gave away my most recent copy to the 12-year-old daughter of a friend. I will be getting another copy for myself as soon as I can."

—John B. Cosgrave, St. Patrick's College, Dublin, Ireland

"After reading *From Zero to Infinity*, I was hooked. This book discussed many beautiful ideas and facts about the integers and posed several interesting problems. I tried to solve them. I failed. I tried to construct counter examples. I made my own conjectures and proved some related results. By the time I graduated from high school I had filled two 2-inch notebooks with my own ideas, results, and calculations."

—Nathaniel Dean, Mathematics Department,
Bell Communications Research

"Constance Reid's book *From Zero to* [illegible] Japanese, and I found it when I w[illegible] was really impressed by Reid's [illegible]. Inspired by it, I even tried to s[illegible]. Now I am working in analytic numbe[illegible] of the reasons for my choice is a sentence [illegible] Analytic number theory is said to be technically the [illegible]ult in the whole of mathematics.' "

—Kohji Matsumoto, student at Nagoya University

"Yesterday, aboard a flight from Denver, I had a nice conversation with a gentleman, John Moulter. When I learned that Mr. Moulter is a retired Los Angeles mathematics teacher, I mentioned that I had the good fortune to know Connie Reid. Mr. Moulter abruptly demanded, 'Do you mean Constance R-E-I-D?' When I answered affirmatively, he told me that you had changed his life. About fifty years ago, then a high-school history teacher, he picked up a copy of *Esquire* magazine in a barbershop containing a review of *From Zero to Infinity*. The review prompted him to buy your book. And your book inspired him to switch from teaching history to teaching mathematics!"

—Letter from a friend with appended note
from John Moulter: "Thank you. Thank you."

"I was the sort of child who always carried a book wherever he went. In fifth and sixth grade that book was most frequently *From Zero to Infinity*. I was indeed born to be a mathematician, but *From Zero to Infinity* helped me to realize that I was part of a community of number-people. There can be few pleasures more satisfying than having the chance, as an adult, to help bring one's favorite childhood book back into print."

—Bruce Reznick, University of Illinois Urbana

"I want to thank you for having written such a wonderful book. It was pitched on just the right level for a young teenager but, more to the point, it expressed the right mix of beauty and wonder. I just had to learn more. I very much believe that this small book, which still occupies an important place in my personal library, enriched my life immeasurably. It is very rare that we find what we really want to do in life, and I am very grateful that your book led me in the right direction."

—Hugh Williams, University of Calgary, Canada,
author of *Edouard Lucas and Primality Testing*

· FROM ZERO TO INFINITY ·

· FIFTIETH ANNIVERSARY EDITION ·

From Zero to Infinity

What Makes Numbers Interesting

Constance Reid

A K Peters, Ltd.
Wellesley, Massachusetts

Editorial, Sales, and Customer Service Office

A K Peters, Ltd.
888 Worcester Street, Suite 230
Wellesley, MA 02482
www.akpeters.com

First published in 1955 by Thomas Y. Crowell Co.
Second edition, 1960 by Thomas Y. Crowell Co.
Third edition, 1964 by Thomas Y. Crowell Co.
Fourth edition, 1992 by The Mathematical Association of America.
Fifth edition, 2006

Library of Congress Cataloging-in-Publication Data

Reid, Constance.
From zero to infinity : what makes numbers interesting /
Constance Reid.– 5th ed.
p. cm.
ISBN-13: 978-1-56881-273-1 (alk. paper)
ISBN-10: 1-56881-273-6 (alk. paper)
1. Numerals. 2. Number theory. I. Title.
QA93.R42 2006.
510--dc22 2005027860

Printed in India
10 09 08 07 10 9 8 7 6 5 4 3 2

· IN MEMORY OF ·

Julia Bowman Robinson
1919–1985

and

Raphael Mitchel Robinson
1911–1995

· CONTENTS ·

$0,\ 1,\ 2,\ 3,\ 4,\ 5,\ 6,\ 7,\ 8,\ 9,\ \ldots$

The natural numbers, which are the primary subject of this book, do not end with the digits with which we represent them. They continue indefinitely—as the three dots indicate—to infinity. And they are all interesting: for if there were any uninteresting numbers, there would of necessity be a smallest uninteresting number and it, for that reason alone, would be very interesting.

· ACKNOWLEDGMENTS ·

Throughout the half century during which *From Zero to Infinity* has been in print, it has had a number of different publishers. As its author I would like to express my special gratitude to four of them.

First, to Dennis Flanagan, the longtime editor of *Scientific American,* who accepted an article on "Perfect Numbers" from a freelancing housewife who was not even a mathematician.

Second, to Robert L. Crowell, who read her article in *Scientific American,* saw its possibilities for a book, and shepherded it and its author through three editions.

Third, to Donald J. Albers, publications director of the Mathematical Association of America, who, after I had retrieved the copyright from the last in the series of publishing companies that had come into possession of Crowell books, published a fourth edition of *From Zero to Infinity* under the imprint of the MAA.

Fourth, to Klaus Peters, the president of A K Peters, Ltd. I am particularly happy that Klaus will be republishing my first "mathematical" book, because he was also the publisher, as mathematics editor of Springer-Verlag, who in 1969 accepted for publication my life of David Hilbert and thus opened up to me a new field of mathematical writing—the writing of mathematical lives.

—Constance Reid

· AUTHOR'S NOTE ·

There is a story behind the publication of this fiftieth anniversary edition of *From Zero to Infinity*.

It begins with a phone call from my sister, Julia Robinson, on the morning of January 31, 1952. She has called to tell me of an exciting event that occurred the night before at the Institute for Numerical Analysis on the UCLA campus, where the National Bureau of Standards has located its Western Automatic Computer—the SWAC.

Julia tells me that a program by her husband, Raphael Robinson, had turned up the first new "perfect numbers" in seventy-five years—not one but two of them. (I learn only later from others that Raphael had at this point never seen the SWAC and had programmed entirely from a copy of the manual.) Julia explains the problem simply: *perfect numbers*—the name itself is intriguing—are numbers like 6 that are the sum of all their divisors except themselves: $6 = 1 + 2 + 3$. Then she tells me there is a particular form of prime necessary for the formation of such numbers, the amount of calculation involved in determining their primality, the enormousness of such primes. For me the whole thing is fascinating. I decide to write an article about the discovery of new perfect numbers.

I am lucky to be able to interview Dick Lehmer, the Director of the SWAC, while he and his wife Emma are visiting in Berkeley. It is Emma who suggests that I send my article to *Scientific American.* If you look up the March 1953 issue you will see a photo of the SWAC and be able to read a fairly detailed description of Raphael's program and how it was fed into the computer.

Of course a subscriber later wrote to Dennis Flanagan, the editor, to complain that when he read an article in *Scientific American* he expected the author to be a Ph.D. But my not being a Ph.D. did not seem to have concerned Dennis Flanagan anymore than it concerned the publisher Robert L. Crowell. After reading my article, Mr. Crowell immediately wrote to ask if I would be interested in writing a small book on numbers that he could pair with a book on the alphabet. Even I found the combination a bit incongruous, but it gave me an idea. The title of Mr. Crowell's book, already in print, was *Twenty-six Letters.* I would write a book about the ten digits; and because I had found what Julia had told me to be so interesting, I would call it *What Makes Numbers Interesting.*

I consulted Julia and Raphael. Robert Crowell's proposal was something of a joke to them—Constance writing a book about "mathematics"—but they thought I could do what I proposed. They would, they promised, read the manuscript before it went to the publisher. Otherwise I was on my own.

I promptly sent off a proposal and a sample chapter to Robert Crowell. He replied that although he had found my sample chapter on zero "pretty tough sledding," he was enclosing a contract.

Today I really don't know how I managed to write the book that I wrote. But I learned a great amount in the course of doing so, and I found what I was learning so very interesting that I didn't see how it could fail to interest others.

The book was finished in a little over a year,

Then came a problem. The sales department flatly vetoed my proposed title—*What Makes Numbers Interesting*. The word *interesting* bothered them. Nobody would buy a book about things that were described as "interesting".

Mr. Crowell agreed. Could Mrs. Reid come up with another title?

I submitted a dozen or so, none of which I liked. The sales department simply loved the one that I disliked the most—*From Zero to Infinity*. My reasons for disliking it were the following. First, in ascending order of importance, it was similar to the title of a then very popular novel, *From Here To Eternity*. (A later reviewer noted that the title sounded as if the book was a novel.) Second, it was too similar to George Gamow's *One, Two, Three . . . Infinity* (although Gamow had begun with the number 1 while I had begun with the number 0). My real objection to the proposed title, however, was that I had not written anything in my book about the theory of the infinite.

So I added three dots after the chapter on nine to indicate that the natural numbers continued "to infinity," and held out for my original title as a subtitle. It is still there fifty years later—*What Makes Numbers Interesting*—along with a neat little proof that there is no such thing as an *uninteresting* number.

The book did quite well. It was recommended for teachers and libraries and selected by science book clubs, even described as doing for number what George R. Stewart's *Storm* had done for weather: "breathing life into a seemingly lifeless body." (Stewart had been the first to give girls' names to storms.)

The Russians put Sputnik into space in 1957. Americans became almost hysterical about the possibility of falling behind in mathematics and science. Mr. Crowell announced that it was time for a second edition. At this point the change in title, to which I had so strenuously objected, paid off. Since there must be a new chapter for a new edition, it would be a chapter on the theory of infinite sets.

Four years later Crowell wanted still another edition—and of course another chapter. What could follow "Infinity"?

Here Raphael came to the rescue, proposing a chapter on the base of the natural logarithm. As he reasonably pointed out, it was only with *e* that mathematicians had finally been able to establish—by mathematical proof—the distribution in the large of the prime numbers—in short, to prove the Prime Number Theorem. For this reason, in his opinion, *e* was not at all out of place in a book on the natural numbers.

Some time later, after the third edition of my book, Robert Crowell had to face the fact that none of his four sons wanted to carry on the publishing firm that had been founded by their great grandfather. The firm was sold. It became Lippincott, then Harper and Row, then, as I recall, HarperCollins. Eventually royalties were so meager that the book seemed essentially out of print. I asked for the copyright to be returned and HarperCollins agreed, retaining only the rights to a Japanese translation.

The fourth edition of *From Zero to Infinity* was published in 1992 by The Mathematical Association of America with an autobiographical author's note instead of still another chapter.

You are now reading the Fiftieth Anniversary Edition of *From Zero to Infinity,* which is being published by A K Peters, Ltd. It is the hope of both the publisher and the author that the story that began with the discovery of the first new perfect numbers in 75 years will continue and that, through this new edition, the book will continue to reach out to new generations of young people, some of whom may be inspired, as others have been, to become mathematicians—and all will gain a glimpse of what has made the natural numbers so eternally interesting.

· ZERO ·

Zero is the first of ten symbols—the digits—with which we are able to represent any of an infinitude of numbers. Zero is also the first of the numbers that we must represent. Yet zero, first of the digits, was the last to be invented; and zero, first of the numbers, was the last to be discovered.

These two events, the invention and the discovery of zero, tardy as they were in the history of number, did not occur at the same time. The invention of zero preceded its discovery by centuries.

At the time of the birth of Christ, the idea of zero as symbol had occurred only among the Babylonians and had vanished with them. The idea of zero as number had not occurred to anyone. The problem of writing down numbers without using a different symbol for each one had been met in very much the same way by all the other great civilizations. The Egyptians had used appropriate pictures; the Greeks, the letters of their alphabet; the Romans, the few simple lines that we see so often on cornerstones; but all had grouped the numbers so that the same symbols could be used over and over. It was possible to write down numbers, but it was not possible to write them down in a way that they could be easily handled in even the simplest processes of arithmetic. Anyone who has tried to multiply Roman numerals will have no difficulty in understanding why the Roman, when he had a problem in arithmetic, turned his back on the Vs and Xs and Cs and Ms of the written numbers and obtained his answer with the beads on a counting board. Egyptian and Greek did the same thing. Yet it never seemed to occur to any of them that in these same beads was the secret of the most efficient method of number representation that the world was to develop.

The counting board, although it took various forms and names in various civilizations, was basically a frame divided into parallel columns. Each column had the value of a power of ten, the number of times that a particular power occurred in a total being represented by markers of some sort, usually beads. All the beads were identical in appearance and all stood for one unit. The value of the unit, however, varied with the column. A bead in the first column had the value of 1 (10^0), in the second column of 10 (10^1), in the third of 100 (10^2) and so on. For this reason the uncertain life of a favorite at the court of a tyrant was sometimes compared to that of a marker on the counting board "which signifies now much, now little."

Numbers that the Romans represented in writing as CCXXXIV (234) and CDXXIII (423) were easily distinguishable on the board.

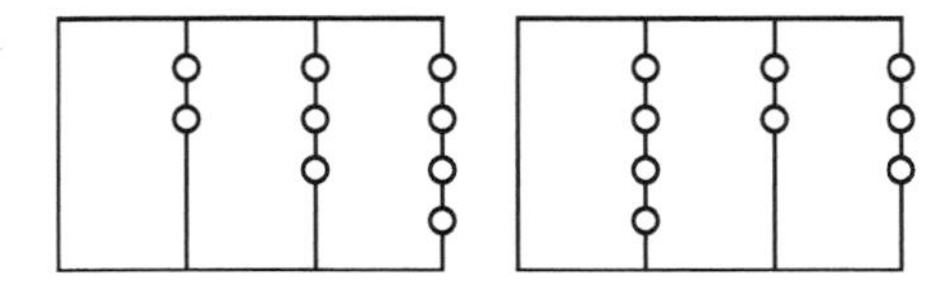

Today we are immediately struck by the resemblance between this ancient method of representing numbers on the counting board and our present method of representing them in writing. Instead of nine beads, we use nine different symbols to represent the total of beads in a column and a tenth symbol to indicate when a column is empty. The ordering of these ten symbols, which we call the digits, tells us exactly the same thing the beads do: 234 tells us two 100s, three 10s, four 1s, while 423 tells us four 100s, two 10s and three 1s.

In short, modern *positional* notation, where each digit has a varying value depending upon its position in the representation of a number, is simply the notation of the counting board made permanent. All that is needed to transfer a number from the board to paper is ten different symbols; for there can be only one of ten possible totals in a column: one, two, three, four, five, six, seven, eight or nine beads, or no beads at all. The column can be empty, and the tenth symbol must of necessity be a symbol for such an empty column. Otherwise it would be impossible to distinguish among different numbers from the counting board.

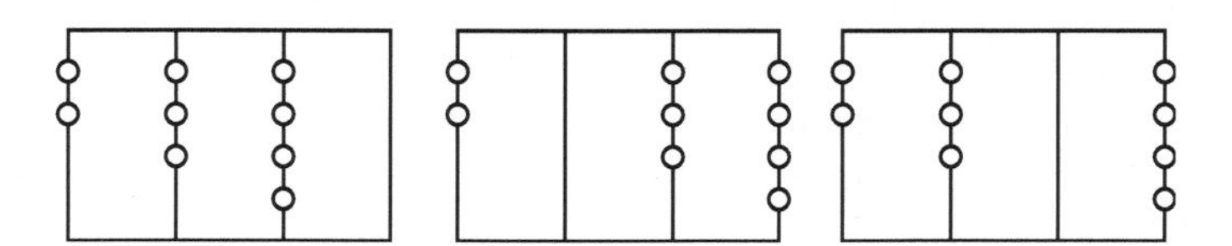

Without such a symbol the above examples would all on paper be the same: 234. With a symbol they are easily distinguishable as 2,340, 2,034 and 2,304.

It would seem that the first time anyone wanted to record a number obtained on the counting board, he would automatically have put down a symbol of some sort—a dash, a dot or a circle—for that empty column, which we today represent by 0. But in thousands of years, nobody did.

Not Pythagoras.

Not Euclid.

Not Archimedes.

For the great mystery of zero is that it escaped even the Greeks, except for their astronomers.

It is difficult, if not impossible, as the reader of this book will soon discover, to write about numbers without writing about the Greeks. The respect in which mathematicians hold these ancient "contemporaries" was expressed by England's G. H. Hardy (1877–1947) when he wrote: "Oriental mathematics may be an interesting curiosity, but Greek mathematics is the real thing. As Littlewood[1] said to me once, 'The Greeks are not clever schoolboys or scholarship candidates, but *fellows at another college*.'"

That zero, or nothing, was not recognized as a number by the Greeks is more than curious. They were the first people to be interested in numbers solely because numbers are interesting, and they left to number theory some questions that remain unanswered to this day. They were concerned, however, with learning the secrets of numbers, not with using them. They looked at the numbers through the lens of geometry, and this may be one reason that the idea of zero as a number escaped them. Moreover, although much of number theory has no need for a zero, without a zero reckoning is impossibly hobbled. The great Greek mathematicians, pondering the interesting numbers, considered that reckoning was an occupation for slaves and left it to them.

It was India that gave us our zero, and with it a practical system of arithmetical notation. Sometime in the early

[1]J. E. Littlewood (1885–1977).

centuries of the Christian era an unknown Hindu who wanted to keep in permanent form the answer on his counting board put down a symbol of his own invention, a dot he called *sunya,* to indicate a column in which there were no beads.

Thus, after all the others, came zero, the first of digits.

It has been pointed out by some that the invention of a symbol for nothing, the void, was one for which his philosophy and religion had peculiarly prepared the Hindu. But it must be understood that the dot *sunya* which the Hindu created was not the number zero. It was merely a mechanical device to indicate an empty space, and that was what the word itself meant—empty. The Indians still use the same word and symbol for the unknown in an equation—what we usually refer to as x—the reason being that until a space is filled with the proper number it is considered empty.

With *sunya,* the symbol zero had been invented; but the number zero was yet to be discovered. In the meantime, the new Indian notation made its way to Europe through the Arabs as "Arabic" notation. Immensely superior as it was, it was not immediately accepted. In 1300 the use of the new numerals was forbidden in commercial papers because it was thought that they could be forged more easily than the Roman numerals. Merchants recognized their usefulness while the more conservative class of the universities hung onto the numerals of the Romans and the system of the counting board. Not until 1600 were the new numerals accepted all over Europe.

Everyone recognized that the revolutionary thing about the notation was the inclusion of the dot—*sifr,* as it was called in Arabic—to represent the empty column. The whole new system came to be identified by the name of this one symbol, and that is how the word *cipher,* in addition to standing for zero, came also to stand for any of the digits and the verb *to cipher*, for *to calculate*. (*Zero* came later from the French and the Italian.) But *sifr*, like *sunya*, was still a symbol for an empty column, not a number.

Even today, although we use the symbol 0 constantly, we do not always think of it as a number. On a typewriter keyboard or a telephone dial, we still list it with the other digits but place it after 9. Since in value it does not exceed 9, it is obviously there as a symbol and not as a number.

This should not surprise us, for zero is the one digit that we do not commonly use as a number. If the reader will answer the few questions below he will discover for himself that he is much more efficient in handling zero the symbol than in handling zero the number. The symbol is the zero he knows; for it is a curious fact that positional arithmetic, which depends for its existence upon the symbol zero, often gets along very well without the number zero.

· UNDERSTANDING ZERO ·

Zero as a Symbol	*Zero as a Number*
$1 + 10 =$	$1 + 0 =$
$10 + 1 =$	$0 + 1 =$
$1 - 10 =$	$1 - 0 =$
$10 - 1 =$	$0 - 1 =$
$1 \times 10 =$	$1 \times 0 =$
$10 \times 1 =$	$0 \times 1 =$
$10 \times 10 =$	$0 \times 0 =$
$10 \div 1 =$	$0 \div 1 =$
$1 \div 10 =$	$1 \div 0 =$
$10 \div 10 =$	$0 \div 0 =$

· ANSWERS ·

As a Symbol: 11, 11, −9, 9, 10, 10, 100, 10, 1/10, 1.

As a Number: 1, 1, 1, −1, 0, 0, 0, 0, impossible, indeterminate.

Centuries after *sunya* had been invented as a symbol for the empty column on the counting board, people were still fumbling toward the mastery of zero as a number that could be added, subtracted, multiplied and divided like the other numbers. To the scholar of today, poring over ancient mathematical papers, the test of mastery is always the same. Addition, subtraction, even multiplication with zero seem to have caused relatively little trouble. Always it is the handling of division of and by zero that shows us today whether a person really understood the curious new number. The problems that caused the trouble were similar to the last three in our little test (probably the same ones that caused the reader trouble).

$$0 \div 1 = ?$$

The fractional expression $0/1$, which is just another way of expressing the division, is mathematically meaningful. Zero can be divided by any other number; in this it is unique among the numbers. (In number theory, one number is considered to "divide" another only when the answer obtained is a whole number.) No matter what number is multiplied by zero, the answer is always the same—zero. Since $0 \times 1 = 0$, $0 \div 1 = 0$. No matter what number is divided into zero, the answer is always the same—zero.

$$1 \div 0 = ?$$

The expression $1/0$ is not, on the other hand, mathematically meaningful. Zero cannot divide any number except itself, not even as the denominator in a fractional expression. In this, and in the fact that it can be divided by all numbers, it is unique. The reason that $1/0$ is a meaningless expression is the same reason that $0/1$ is a meaningful one.

No matter what number is multiplied by zero, the answer is always zero. A division, however, indicates that some number (the quotient) when multiplied by another (the divisor) will produce the number being divided. If there is an answer to the problem $1/0$, or a value for the expression $1/0$, it would have to be such a number that multiplied by zero would produce one. But we have already stated that *any* number multiplied by zero can produce only zero. It follows, therefore, that we cannot divide one (or any other number) by zero.

$$0 \div 0 = ?$$

The expression $0/0$ is neither mathematically meaningful nor meaningless. It is *indeterminate.* Zero can be divided by itself, but there is no way of determining the value of the answer. Since any number multiplied by zero produces zero, zero divided by zero can yield any number. Zero divided by zero can equal zero, since $0 \times 0 = 0$, but it can also equal one, since $0 \times 1 = 0$, and two, since $0 \times 2 = 0$, and so on and on. Zero has always been a favorite in a field of insult best described as "mathematical invective." An example from a newspaper is "a lousy nothing divided by nothing." This is mathematically a less definite insult than was intended.

The three terms we have been using—*meaningful, meaningless* and *indeterminate*—can be made even clearer by a comparison. An indicated operation of division is said to be mathematically meaningful only if it stands for a specific value that can be obtained by performing the operation. It may be compared to a title used to identify specifically a person not named. The President of the United States, for instance, is such a title. Generally when we use it we are referring to a certain person as specifically as if we had named him. In a similar way the expression 0/1 (or $0 \div 1$) refers to a specific value: 0. It cannot stand for another value anymore than 10/1 can stand for a value other than 10.

An indicated operation of division, on the other hand, is mathematically meaningless when it cannot possibly have any value. In the same way a title may be meaningless. The King of the United States is such a title. The expression 1/0 (or $1 \div 0$) is meaningless because one cannot be divided by zero; therefore, the expression stands for no value. (*No value* is not at all the same thing as zero.)

The expression 0/0 (or $0 \div 0$) is meaningless in a quite different sense. It is like the title the United States Senator, which is meaningless for identification unless the context in which it is used specifies which of the one hundred senators is meant by it. The choice with the expression 0/0 is much greater than one hundred. It can have any numerical value we choose to give it, since any number multiplied by zero produces zero. The expression 0/0 is meaningless only because it can mean anything. Mathematicians say, more technically, that it is *indeterminate*, and it took them centuries to realize that it is. Only then had they finally mastered zero the number.

To understand the special significance of zero among the numbers, we must examine what are known as the integers. When the integers are arranged in order, the positive numbers, which we might say count things present, extend indefinitely to the right; the negative numbers, which count things absent, extend indefinitely to the left. This is an arrangement we are familiar with on the thermometer, the positive numbers being the degrees "above" zero; the negative, those "below."

$$\ldots, -5, -4, -3, -2, -1, 0, +1, +2, +3, +4, +5, \ldots$$

In this arrangement of negative and positive integers, every consecutive pair must be the same distance apart as every other pair. Such regularity of spacing is the essence of the integers: -1 is the same distance from -2 that $+1$ is from $+2$ and also $+2$ is from $+3$. But this regularity can be maintained only if zero is included as one of the integers. Without zero the distance between -1 and $+1$ is twice the distance between any other pair. Obviously then -1 and $+1$ are not consecutive: 0 is the number between them.

In the Christian accounting of time, unlike on the scale above, zero marks not a number but a point. A problem in degrees of temperature, therefore, yields quite a different answer from a similar problem in years. If the temperature is 5° below zero in the morning and rises 8° during the day, it is then 3° above zero. But a child born on the first day of January, 5 B.C., will not be eight years old until 4 A.D. The reason for the difference in the answers to these two seemingly identical problems is clear when we place the scale of temperature against the scale of time:

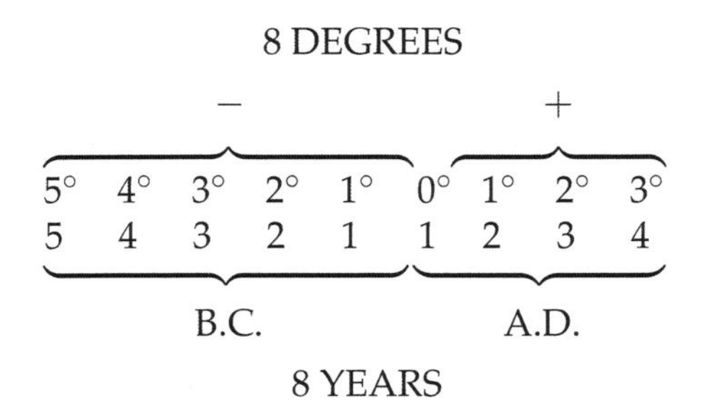

This difference was the cause of a major howler in the scholarly world in 1930. The celebration of the 2,000th anniversary of the birth of the poet Virgil was in full swing when a mathematical killjoy pointed out that there having been no year zero, the poet (born in 70 B.C.) would not have been 2,000 years old until 1931. The scholars, who should have known better, were performing on a scale on which zero is not a number, a mathematical process that works as it does because zero is a number. A similar thing happened when January 1, 2000, was celebrated as the beginning of the new millennium, which actually began on January 1, 2001.

Among the whole numbers, or integers, zero is unique, being neither negative nor positive. Although we use all the integers in computation, it is those after zero that we usually think of as "the numbers." (As late as the twelfth century the Indian mathematician Bhaskara gave $x = 50$ and $x = -5$ as the roots of the equation $x^2 - 45x = 250$, but cautioned, "The second value in this case is not to be taken, for it is inadequate; people do not approve of negative roots.") We even call the numbers that follow zero the *natural* numbers, although it may be argued whether they are in fact more natural than any other numbers. They are the numbers we count with, and that seems to us the natural thing to do with numbers. We do not think of zero as one of them because it does not seem at all natural to most of us to "count" with "nothing." It is, however, becoming more natural as more and more people learn to use computers. As far as the computer is concerned, zero is a number. There is no question about that. Not only does 0 precede 1 on the numerical pad—as it should—but in programming even such everyday things as house and car payments, what we usually think of as the "first" year must be treated, not as year one, but as year zero.

Zero, unlike the negative numbers, is logically at home with the so-called natural numbers. For zero answers the same great question that all the counting numbers answer, and it answers it in exactly the same way. The question is simply, *How many*?

> How many people are there in the room where you are reading this book?
>
> How many elephants are there in the room where you are reading this book?

The answer to the first question is at least one, maybe two or three; but the answer to the second is quite probably zero. The number of elephants in the room is zero. Zero is a number just like one and two and three.

But if zero is a number, the reader may well ask, just what is a number anyway?

Certainly a number is an abstraction, a recognition of the fact that collections may have something in common even though the elements of the collections have in common nothing whatsoever. There is a similarity between two mountains and two birds even though birds and mountains are not similar, and this similarity they share with two of anything. While this may seem obvious to us, it was not to our ancestors. They recognized the difference between one pheasant and two, one day and two but, as Bertrand Russell (1872–1970) has pointed out, "It must have required many ages to discover that a brace of pheasants and a couple of days were both instances of the number two."

What was discovered, mathematically speaking, was that the number two is the common property of all sets containing a pair. It does not matter whether a set contains people, elephants, flies or mountains, or completely different objects; it shares with all other sets that contain a pair the number two.

When we say that one, two and three are numbers, we mean that one is the number of all those sets that contain a single member; two is the number of the sets that contain a pair; three is the number of the sets that contain a triple. Since there is no end to the possibilities of what these sets may contain, we say that they are infinite.

There is also a set of zero, comparable to these others. This is the set that contains no people, no elephants, no flies, no mountains. In other words, the empty set. In the same way that one, two and three are the numbers of the sets of one, two and three respectively, zero is the number of the empty set. There is, however, a difference between the set of zero and the other sets that has nothing to do with the difference in the number of members. While all the other numbers represent an infinite number of sets, zero represents only one, the empty set. Whether it is empty of men, elephants, flies or mountains does not matter; it is the same set—and there is only one.

It is things like this that make zero a very interesting number among an infinitude of interesting numbers. Each of the natural numbers is, of course, unique: two is not three, and three is not four, and four is not five—or any other number. But the uniqueness of zero is more general than that of the other numbers, more significant for that reason, and therefore more interesting.

> Zero is the only number that can be divided by every other number.
>
> Zero is the only number that can divide no other number.

Because of these two characteristics, zero is almost invariably a "special case" among the numbers, and we shall find many examples of its "specialness" in the pages to follow. Zero is enough like all the other natural numbers to be one of them, but enough different to be a very interesting number: the last, and the first, of the digits.

· A PROBLEM ·

The digits can be arranged in various ways. In this chapter we have mentioned two. In one arrangement 0 follows 9 as a symbol; in the other, much less common, 0 the number comes before 1. But usually, even in fun, 0 ends up as the last of the digits. The basis for the arrangement below is one that many mathematicians have great difficulty in perceiving.

8 5 4 9 1 7 6 3 2 0

· ANSWER ·

The digits are arranged in alphabetical order. Secretaries usually outwit mathematicians on this one. A variation is to arrange the digits in a language other than English.

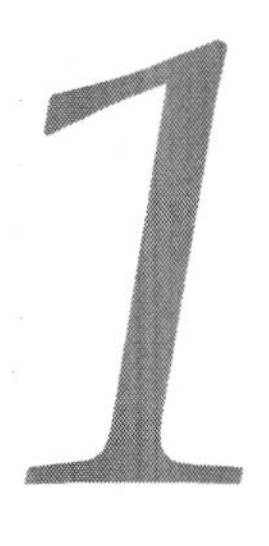

· ONE ·

We are all familiar with the behavior of the number one in the ordinary processes of arithmetic. It does not surprise us as does the behavior of zero. In fact, it is so simple that we generally dismiss it as trivial. We do not even bother to learn the "ones" in school, so obvious is it to us that any number when multiplied by one yields a product that is itself and when divided, a quotient that is itself. Yet these simple characteristics of the number one have the greatest implications for the study of numbers.

The very first idea of number comes with the recognition that there is a difference between one and more than one. A child grasps this distinction when he is about eighteen months old. "He likes to assemble the many cubes into a pile," writes Arnold Gesell in *The First Five Years of Life*, "or to disperse the pile into the many cubes. . . . In comparison, the one-year-old is single- and serial-minded."

Presumably man recognized the distinction proportionately early in his history. Either there is one wolf or there are many around the fire, one river or more than one between this camp and the next, one star in the early evening sky or many when the campfire dies down. We begin with two number words, but only the number one. Nevertheless it is possible for us to count in a fashion, and quite accurately, with this number. We look up suddenly from the campfire and see "many" wolves—more than one. There are actually two wolves, but we have no word for two, so we say there are many. How many? We try to think of something as "many" as the wolves and come up with another pair with which we are familiar. We announce that there are as many wolves as a bird has wings.

This method of communicating the exact number of wolves, distinguishing among the many meanings of "many," need not stop with a pair. We can find other sets with which we are familiar and against which we can match the number of wolves, one to one. The wings of a bird may be followed by the leaves of a clover, the legs of an animal, the fingers of a hand. We are then able to "count" any number of wolves from one to five although we still have no number other than the number one.

We look around for what is logically the next set, a set that contains one more member than our hand has fingers. It is not so easy to find an immediate set of six in nature. So instead of using another completely new set for counting one more wolf, we add to the set of fingers on one hand a finger from the other. This is a good practical idea because now, for our next set, when the number of wolves increases by still one more, we can add another finger and can continue in this manner until we have used all the fingers of both hands to count ten wolves.

But the wolves keep coming. What can we do with a many that is more than the fingers on both our hands? We could of course start upon our toes, and some people did; but we decide to reuse our fingers. For one more wolf, we put up both our hands and then a finger by itself. We have now started, inexorably, on our way to infinity. We will never get there, but we will never have to stop along the way and say that we can't go any farther. For no matter how many wolves we have "counted" and how many fingers we have used in the process, we can always lift one more and count one more wolf.

What then has been our achievement?

Simply that we have constructed, with no other number concept except that of the number one, the infinite set of natural numbers:

$$
\begin{aligned}
&1, \\
&1+1, \\
&1+1+1, \\
&1+1+1+1, \\
&\ldots .
\end{aligned}
$$

These *are* the natural numbers: the foundation upon which has been erected the beautifully complex edifice that is the theory of numbers.

The fact that one generates all the other numbers by successive additions of itself has always, even after it was no longer the only number, given it a special significance. The Greeks had a hard time defining one because it was the means by which they defined all the other numbers. *Could the maker of numbers be itself a number?* they asked themselves. They decided it could not. (As Aristotle reasonably observed, the measure is not measures but *the measure*.) So instead they defined one as the beginning, or principle, of number. It was so completely set apart from the other numbers that it was not considered the first odd number (that was three), but rather the great Even-Odd because when added to odd numbers it produces even and when added to even, odd. One was not a number but Number with a capital "N." It was considered to contain within itself, layer by layer like an onion, all the other numbers.

The onion simile is not farfetched. Joseph T. Shipley, in his *Dictionary of Word Origins,* remarks: "Those that have, with intended humor, transposed the saying 'In *union* there is strength' to 'In *onion* there is strength' in all probability did not know that in *onion* there is *union.* With the same vowel change as in one, from L. *unus, one, onion* is from L. *unio, union, unity,* from *unus.* The idea is that the many layers make but *one* sphere... . The *onion* has been used as a symbol, in that, far as you may peel, you never reach the core."

This great reversal of *e pluribus unum* that is embodied in the number one has always given it first place among the numbers in religion. During the Middle Ages, when mysticism flourished while mathematics languished, the number one represented God the Creator, the First Cause, the Prime Mover. The other numbers were considered more imperfect in direct proportion as they receded from one. Two, as the first number so receding, signified sin, which deviates from the first good. Fortunately for the larger numbers, there were ways by which they could be reduced to the digits so that they weren't completely beyond salvation.

The characteristics of the number one that give it so much nonmathematical significance are the same that make it mathematically interesting—and the same that make its behavior so obvious and hence so seemingly trivial in the ordinary processes of arithmetic:

> One is the only number that divides every number.

One is the only number that no other number divides.

Among the infinitude of natural numbers, each in its own way unique and yet in many ways very like the others, there is no other number at all like one. The only number ever linked with it is its antithesis—zero. While one divides all the numbers, zero divides none; while one is divided by none of the other numbers, zero is divided by all of them. Among the numbers they are both "special cases."

The behavior of one that seems so trivial to us in multiplication and division is the direct result of the ability of one to generate all the other numbers by successive additions of itself. One is the unit out of which the other numbers are built. Do not let the grammar-school obviousness of this fact deceive you, for it is the most important single fact in all the theory of numbers. When we are trying to wrest the secrets of their relationships from the numbers, the fact that one divides *all* the numbers is a most valuable weapon. It is in a sense the weapon with which we start. Our next follows from it—the fact that every number is also divisible by itself.

Given the infinite set of natural numbers, each differing from its predecessor by one, the theory of numbers extends a challenge. What can be learned about the numbers besides these two easily ascertained facts—that every number is divisible by one and that every number is divisible by itself?

The first step toward the understanding of any group, including the numbers, is the classification of its members into mutually exclusive subgroups. At first thought it might not seem that the two facts we have been given would provide a basis for such a classification. It was not, indeed, the first thought of man that they did. The most ancient classification of numbers into groups was on the basis of their divisibility by the number two. The numbers that are exactly divisible by two were called *even* and those that leave a remainder of one when divided by two were called *odd*. All numbers belong in one of these groups, and no number in both. The even-odd classification seemed so basic to the Greeks that they thought of it like the great distinction between the two kinds of human beings. The even numbers they saw as "ephemeral," hence female; the odd, "indissoluble, masculine, partaking of celestial nature." But even-odd, based on divisibility by two, is not nearly so significant a classification of the numbers as one that is based on their general divisibility.

We have already made two statements about the general divisibility of numbers, and we can add to these two more, which we arrive at after examining the first few numbers and their divisors.

> Some numbers, like two, three, five and seven, are divisible only by themselves and one.
>
> Some numbers—four, six, eight and nine, for example—are divisible as well by some number other than themselves and one.

Here is a basis for a classification of numbers into two groups that has produced enough mathematics to fill most of the bulky first volume of L. E. Dickson's three-volume definitive history of the theory of numbers. The numbers in the first group, divisible only by themselves and one, are commonly called the *prime* (or first) numbers. Since it can be proved quite simply that all the numbers in the second group, divisible by some other number in addition to themselves and one, are composed of prime numbers, these are known as *composite* numbers.

(If a number n is composite, it has by definition divisors between one and n. If m is the least of these divisors, it must be prime because otherwise [if it is divisible by a number other than itself and one] it cannot be the least divisor of n. Continuing in this way, we can reduce all the divisors of n to primes, thus proving that every composite number can be produced by primes.)

We saw a few pages back how it is possible for us to represent all the numbers after zero by successive additions of one. Now we see that after zero and one (which, befitting their status as special cases, are neither prime nor composite[1]) we can also represent all the numbers by primes or combinations of primes:

$$
\begin{array}{lll}
1+1 & = 2 & \text{(prime)} \\
1+1+1 & = 3 & \text{(prime)} \\
1+1+1+1 & = 2 \times 2 & \text{(composite)} \\
1+1+1+1+1 & = 5 & \text{(prime)} \\
1+1+1+1+1+1 & = 2 \times 3 & \text{(composite)} \\
1+1+1+1+1+1+1 & = 7 & \text{(prime)} \\
\dots & \dots &
\end{array}
$$

We do not need to be told that the additive representation of numbers in the left-hand column is unique. It is

[1]Zero is not prime because it is divisible by an infinitude of numbers besides itself and one and not composite because, since one of its factors is always itself, it cannot be produced by primes alone. One is technically excluded from the primes because, as we shall see, if it were a prime, the most important theorem about primes would no longer be true. In addition, although one is, like the primes, divisible only by itself and one, it has only one divisor while they have two.

obvious that there can be but a single possible way of expressing any number as the sum of ones. If six is 1 + 1 + 1 + 1 + 1 + 1, it can be nothing else; and its successor among the numbers, whether we call it seven or simply the successor of six, can be nothing but 1 + 1 + 1 + 1 + 1 + 1 + 1.

It is not so obvious that the multiplicative representation of numbers in the right-hand column is also unique. Just as there is but a single way of expressing a number as the sum of ones, there is but a single way of expressing it as the product of primes:

> $6 = 1 + 1 + 1 + 1 + 1 + 1$, and nothing else as the sum of ones;
>
> $6 = 2 \times 3$, and nothing else as the product of primes.

There is but a single way (without respect to order) that a number can be produced by primes alone. This is true of any number, no matter how large it is. A number like 17,640, for instance, is the sum of 17,640 ones and its prime factorization is $2 \times 2 \times 2 \times 3 \times 3 \times 5 \times 7 \times 7$. There are no primes except 2, 3, 5 and 7 that will divide 17,640—although, it being such a large number, we might be inclined to think that there would be others. Only one combination of these four prime factors—three 2s, two 3s, one 5, and two 7s—will produce 17,640. Of course, various other numbers also divide it: 6, 10, 14, 21, 35, to name a few, but ultimately these all reduce to primes—and to the primes 2, 3, 5 and 7.

The representation of any number as the product of primes is unique, just as the representation of any number as the sum of ones is unique.

Think for a moment of the significance of this statement. Any number can be a number so large that it has never been written out, a number so large that a man's lifetime would not be long enough to record it on paper (if the paper were long enough); any number can be any number in an infinite number of numbers. Yet, from the information we have just been given, we can make a very significant statement about this most interesting number, any number n.

We can say that n has certain prime factors, which we designate as $p_1, p_2, \ldots, p_r$, and that the prime factorization of n is a unique combination of these factors. The prime p_1 is used so many times, and we indicate this by $p_1^{k_1}$; p_2 is used so many times, and we indicate this by $p_2^{k_2}$; and so on. Just as we can say that $6 = 2 \times 3$ and $17,640 = 2^3 \times 3^2 \times 5 \times 7^2$, we can say of any number n that $n = p_1^{k_1} p_2^{k_2} p_3^{k_3} \cdots p_r^{k_r}$ and *know* that this representation of n as the product of primes is the only possible representation. This knowledge is so important in the study of numbers that the theorem that states it is universally acclaimed the fundamental theorem of arithmetic.[2]

The proof of the theorem, which tells us that the prime factorization of any n is unique, rests upon a secondary mathematical truth (known as a lemma) that a prime that divides the product of two or more numbers will divide at least one of those numbers. In the case of 17,640—the number we used above as an example—this means that 2, 3, 5 and 7, its prime factors, will each divide at least one number in any group of numbers that, multiplied together, produce 17,640. For example: $15 \times 28 \times 42 = 17{,}640$; and, in accordance with the lemma, 2 divides 28 and 42, 3 divides 15 and 42, 5 divides 15, and 7 divides 28 and 42.

[2]The theory of numbers was treated even by C. F. Gauss (1777–1855) as "arithmetic," as in the title of his classic *Disquisitiones Arithmeticae,* or on occasion as "the higher arithmetic."

The proof of the fundamental theorem itself is by *reductio ad absurdam,* a method that has been a favorite with mathematicians since the time of Euclid. It is simply assumed for the purpose of proof that prime factorization is not unique.

Let us say that a number $n = p_1^{k_1} p_2^{k_2} p_3^{k_3} \cdots p_r^{k_r}$ and also $q_1^{l_1} q_2^{l_2} q_3^{l_3} \cdots q_s^{l_s}$, the ps and the qs being separate sets of prime factors. On the basis of the lemma we stated above, we know that since each p divides n, which is also the product of the qs, each p must also divide some q. Since the qs are by definition prime and hence not divisible except by themselves and one, each p must be equal to some q and conversely each q equal to some p. Both sides then must contain the same primes and the prime factorization of n, contrary to our assumption, is unique.

It has been said that this theorem is essential for a systematic science of arithmetic. Certainly arithmeticians consider it so essential that for the sake of it they exclude the number one from the prime numbers. This is because if one is considered a prime, then prime factorization of the numbers is no longer unique. Instead of being able to say that $6 = 2 \times 3$ and nothing else as the product of primes, we would have to admit an infinite number of possible prime factorizations for six and for every other number:

$$6 = 2 \times 3 \times 1,$$
$$6 = 2 \times 3 \times 1 \times 1,$$
$$6 = 2 \times 3 \times 1 \times 1 \times 1,$$
$$\ldots .$$

Because we know by the fundamental theorem that any number can be expressed uniquely in terms of its prime factors, we are able to handle n with much the same ease that we handle a particular number. Because we can do this, we can often prove something true about all numbers that otherwise we would have to prove for each number, one at a time, and would *never* be able to prove for all numbers.

The example usually given in this connection is that of the theorem that states generally which roots of which numbers are irrational. The Greeks discovered and proved that the square root of two is irrational—that it is not expressible as an integer or as a ratio of integers—in other words, a *rational* number—what we commonly refer to as a fraction. They then went on, one at a time, to prove the irrationality of the square roots of three, five, six, seven, eight, ten, eleven, twelve, thirteen, fourteen, fifteen and seventeen—and here stopped. (The omitted numbers are squares of whole numbers.) For all their labor they had proved nothing except that these few numbers out of an infinitude of numbers have irrational square roots. They had proved nothing about their other roots—cube, fourth, fifth and so on through an infinitude of roots for each number. With the fundamental theorem of arithmetic as a tool, however, it is possible to prove simply and directly when any root of any number is irrational.[3]

Proving something about each number—true of one, true of two, true of three, and so on—will never, no matter how high we go, prove with finality that a statement is true of all numbers. The special challenge that the natural numbers offer is to prove that certain things are true of all

[3]The theorem states that the mth root of N is irrational unless N is the mth power of an integer n. In brief, we cannot possibly get a whole number by raising a fraction to any power—no matter how often, for instance, we multiply 3/2 or any other "proper fraction" by itself we will never get a whole number.

numbers without ever having a chance to examine all the numbers individually. The extent to which human beings have met the challenge of the numbers rests upon the fact that one is the unit. One sets the conditions—an infinite set with each member separated from the next by the same unit—and one puts down the weapons:

Every number is divisible by one.
Every number is divisible by itself.

· A QUIZ ·

The whole subject of divisibility is basic to the study of numbers. In the course of this book we shall answer the questions below in more detail, but now the reader may enjoy trying to answer them for himself:

1. Is there a number that has no divisors?
2. How many numbers have only one divisor?
3. How many numbers have only two divisors?
4. How many numbers have an infinite number of divisors?
5. Is there a number not a divisor of any other number?
6. Is there a number that is a divisor of all numbers?
7. How many numbers divide an infinite number of numbers?
8. What is the largest number having no divisors other than itself and one?

9. How many even numbers have only two divisors?

10. After zero, what number has the greatest number of divisors?

· ANSWERS ·

1. No. 2. Only one—one itself, since all other numbers are divisible at least by themselves and one. 3. An infinite number, since a prime has only two divisors and there is an infinite number of primes. 4. Only one—zero, which has as its divisors each and every one of the natural numbers, which are infinite. 5. Yes—zero, which can divide only itself. 6. Yes—one. 7. An infinite number, since every number except zero divides an infinite number of numbers. 8. There is no largest number, since a prime has no divisors besides itself and one and there is no largest prime. 9. Only one—the number two, which is the only prime among the even numbers. 10. There is none, since by multiplying together as many primes as we please we can obtain a number with as many divisors as we please.

· TWO ·

The number two is not generally written as 10, but it can be. For two is 10 in that simple and elegant system of representing numbers known as the binary.

The binary system of numeration has had something of a rags-to-riches history. It is the descendant of man's most primitive method of representing numbers as anything other than the sum of ones. It was the invention of a great mathematician who had high hopes that it might convert the Emperor of China to Christianity. Until the twentieth century it was looked upon as a mere mathematical curiosity. Then, in the middle of that century, it came into its own with the invention of the computer. Its representation of numbers with only two symbols, one and zero, made possible the representation of numbers simply by the condition of a switch, or current, that was either *on* (1) or *off* (0). Almost simultaneously a new word came into the language: *bit* for *binary digit*—a happy choice since it designates the smallest possible amount of information.

The binary system, simple as it appears, is a relatively sophisticated number system, depending as it does on a symbol to indicate an empty column. The earliest base-two system known to man was the pair system. In the pair system there were also just two number symbols, one and two. Three then was one and two, four was two and two, five was two and two and one. This system was probably sugggested to man by the parts of his own body. Eyes, ears, arms, legs—all were in pairs. Although eventually he was to count by tens because he had ten fingers, he started to count by twos, perhaps because he had two hands.

The pair system, primitive though it was—and it was most primitive—met the essential requirements for a workable system of number representation. It was based on a finite numbr of symbols (there were only two), and it could be used to represent any number, no matter how large. It does not seem likely, however, that with the pair system man ever went beyond five.

The binary system is similar to the pair system in that it also requires only two symbols for representation of any number. The difference is that while the pair system represents numbers by twos, the binary system represents them by *powers* of two.

A power of two, as of any number, is simply the result of a self-multiplication. We are all familiar with the squares and the cubes, which are second and third powers respectively, and recognize that the multiplicative process that produces them can continue indefinitely:

$$2^2 = 2 \times 2 = 4,$$
$$2^3 = 2 \times 2 \times 2 = 8,$$
$$2^4 = 2 \times 2 \times 2 \times 2 = 16,$$
$$2^5 = 2 \times 2 \times 2 \times 2 \times 2 = 32,$$
$$\ldots .$$

Such multiplications, even with such a small a number as two, rapidly attain astronomical proportions: 2^2 is only four, but 2^{10} is in the thousands, 2^{20} in the millions and 2^{40} (or the product of forty 2s multiplied together) in the million millions. Obviously, by using powers of two, we can represent numbers much more compactly than in the pair system and hence more efficiently. Consider the expression of even a small number like 30. In the pair system 30 must be $2+2+2+2+2+2+2+2+2+2+2+2+2+2+2$, but in the binary system it is simply $2^4 + 2^3 + 2^2 + 2^1$ $(16 + 8 + 4 + 2)$.

Except for the substitution of powers of two for powers of ten, the binary system works just as the decimal system does:

$$11111 \text{ in the decimal system} = 10^4 + 10^3 + 10^2 + 10^1 + 10^0,$$

$$11111 \text{ in the binary system} = 2^4 + 2^3 + 2^2 + 2^1 + 2^0.$$

The difference in base gives each system an advantage over the other and a corresponding disadvantage. The decimal system, because it has a larger base, is able to represent numbers much more compactly than the binary system. As we see above, 11111 in the decimal system is a number 358 times as large as the decimal number 31, which is represented by 11111 in the binary system. But the binary system, because it has the smaller base, is able to represent numbers with fewer symbols. This means that it requires a smaller multiplication table, an important practical advantage.

In the binary system, the symbol 1 indicates that a particular column contains a power of two; the symbol 0 indicates that it does not. These two symbols are all that are needed, for it is possible to represent any number uniquely as the sum of powers of two. All numbers are either exactly divisible by two or divisible by two with a remainder of one. Since the zeroth power of two is one and the first power is two itself, the powers of two—it can be easily seen—are sufficient to represent any number by the use of only two symbols to indicate the presence or absence of a particular power in the column reserved for it.

The idea that the zeroth power of two is one is hard to accept until we examine the logic behind this apparently illogical statement:

$$\begin{aligned} 2^3 &= 8 = 2 \times 2^2, \\ 2^2 &= 4 = 2 \times 2^1, \\ 2^1 &= 2 = 2 \times 2^0, \\ 2^0 &= 1. \end{aligned}$$

Although we may swear that we have never heard of such a thing, we work every day with this concept. In the familiar decimal system, as in the binary, the first column is reserved for the zeroth power of the base—the ones. Once we accept this, we must then accept the idea that while every other power of zero is zero, the zeroth power of zero is one.

The fact that *all* numbers can be represented with only 1 and 0 fascinated Gottfried Wilhelm von Leibnitz (1646–1716), the inventor of the binary system. Leibnitz was one of the great mathematicians. We have only to read the account of his life in E. T. Bell's *Men of Mathematics* to be awed by the universality of his genius. Writes Bell: "The union in one mind of the highest ability in the two broad, antithetical domains of mathematical thought, the analytical and the combinatorial, or the continuous and the discrete, was without precedent before Leibnitz and without sequent after him."

All mathematics was not enough to occupy this great mind. Leibnitz also had innumerable nonmathematical projects, one of which was the reuniting of the Protestant and Catholic churches. When he invented the binary arithmetic he saw in it, according to another great mathematician, Pierre Simon Laplace (1749–1827), "the image of Creation... . He imagined that Unity represented God, and Zero the void; that the Supreme Being drew all beings from the void, just as unity and zero express all numbers in [the binary] system of numeration." The story is that Leibnitz communicated his idea to the Jesuit who was the president of the Chinese tribunal for mathematics in the hope that it would help convert to Christianity the Emperor of China, who was said to be very fond of the sciences.

The enthusiasm of Leibnitz for the simplicity and elegance of his system was not shared by his fellow mathematicians, for at the time it appeared that the system had nothing more than simplicity and elegance to recommend it. Yet even in Leibnitz's day the principle of representation by powers of two was commonly used by people who would never have recognized a number expressed in the binary system. These people, who knew so little of arithmetic that they did not even try to multiply except by two, had worked out a very neat system of multiplying in this way. In fact, multiplication and division by two were once so commonly used that, as *duplation* and *mediation*, they were considered basic processes of arithmetic along with addition, subtraction, multiplication and division.

Known generally as "peasant multiplication," duplation and mediation worked like this. To multiply 29 by 31, divide 29 by 2 and the answer again by 2 and so on until you have a remainder of 1. Then double 31 the same number of times that you have halved 29, keeping halvings and doublings in parallel columns. Cross out whatever doubling occurs opposite an even halving and add the remaining doublings to obtain your answer:

29	31
14	~~62~~
7	124
3	248
1	496
	899

If the reader will multiply 29 by 31 in the customary way, he will find that he obtains the same answer.

We can understand why the correct answer has been obtained by "peasant multiplication" only if we examine what has been done in terms of the binary system. The successive halvings of 29 have given us a binary representation of that number. All we have to do is to put a 1 after 29 itself (because it is odd), a 1 after each of the other odd halvings, and a 0 after the even halving.

29	1
14	0
7	1
3	1
1	1

We see immediately that 29 in the decimal system is 11101 in the binary. (This is in fact the simplest method of transposing a number from decimal to binary system.) We then recognize the successive doublings of 31 as multiplications by the powers of two in the binary representation of 29.

$$
\begin{array}{rcl}
1 \times 2^0 = 1 & & 1 \times 31 = 31 \\
0 \times 2^1 = 0 & & 0 \times 31 = 0 \\
1 \times 2^2 = 4 & & 4 \times 31 = 124 \\
1 \times 2^3 = 8 & & 8 \times 31 = 248 \\
1 \times 2^4 = \underline{16} & & 16 \times 31 = \underline{496} \\
29 \times 31 = & & 899
\end{array}
$$

The same multiplication performed in the binary system itself looks like this:

$$
\begin{array}{r}
11111 \\
\underline{11101} \\
11111 \\
00000 \\
11111 \\
11111 \\
\underline{11111} \\
1110000011 =
\end{array}
$$

$$2^9 + 2^8 + 2^7 + 2^1 + 2^0 =$$

$$512 + 256 + 128 + 2 + 1 = 899$$

As we have indicated, the simplicity of binary representation—the fact that in it all numbers are merely arrangements of ones and zeros—makes it the ideal system for computers, which are descendants of Leibnitz's calculating machine—a very superior one for its time, since it was able to do multiplication, division, and the extraction of square roots as well as addition and subtraction.

Computers use the binary system, not because they could not be constructed to use the decimal system, but because with decimals they would be much less efficient. Consider a machine working with a number such as the one that for seventy-five years held the honor of being the largest known prime—a number that, in spite of its great size, is divisible only by itself and one. Mathematicians think of it and work with it as $2^{127} - 1$. In the decimal system, however, it is

$$170,141,183,460,469,231,731,687,303,715,884,105,727.$$

In the binary system it is

1111111111111111111111111111111111111111111
1111111111111111111111111111111111111111111
11111111111111111111111111111111111111111.

For decimal representation of the number above, the machine would have to be able to differentiate between ten different possible symbols in each column of the number. For binary representation it need differentiate between only two.

The particular usefulness of binary representation in high-speed computation arises from the fact that the "symbols" for 1 and 0 do not have to be symbols at all. They can be simply an electric impulse for 1, to indicate the presence of a power of two in the column reserved for it, and no impulse for 0, to indicate the absence of a power.

If Leibnitz had invented his binary arithmetic especially for the computing machines of the future instead of for the Emperor of China, he could not have invented a better system. It does not matter to the machines that the binary representation of large numbers takes up a staggering amount of space. In the early days of the computer it did, however, matter to the people who had to enter the numbers, and they were able to use still another number system—the base-sixteen—in which representation is even more compact than in the decimal system. (Today, of course, such an expedient is no longer necessary, since decimals are entered and changed into binary digits by the machine itself.) In the base-sixteen system each column of number representation increases by a power of sixteen instead of by a power of two as in the binary or a power of ten as in the decimal:

$$\begin{aligned}
111 \text{ (base two)} &= 2^2 + 2^1 + 2^0 &&= 7\\
111 \text{ (base ten)} &= 10^2 + 10^1 + 10^0 &&= 111\\
111 \text{ (base sixteen)} &= 16^2 + 16^1 + 16^0 &&= 273
\end{aligned}$$

The base-sixteen system was selected for transposition to and from the binary because its base is a power of two (2^4). Since the binary system is based on two itself, transposition is relatively simple, as can be seen by comparing the first few powers of two in the base-two and in the base-sixteen:

	(in base-two)		(in base-sixteen)	
2^0		1		1
2^1		10		2
2^2		100		4
2^3		1000		8
2^4		10000		10
2^5		100000		20
2^6		1000000		40
2^7		10000000		80
2^8		100000000		100
...		...		...

Although in the example here, a base-sixteen representation looks exactly like a decimal representation, it does not always. For full representation in the base-sixteen, we need six symbols in addition to the ten we use in the decimal system. At one time it was customary to represent 10, 11, 12, 13, 14 and 15 in base-sixteen with the last six letters of the alphabet, *u* standing for 10, *v* for 11, and so on. Later the IBM PC used instead the first six capital letters. Thus *xyz* and *DEF* both stood for $(13 \times 256) + (14 \times 16) + (15 \times 1)$ in base-sixteen. This is the number we recognize in base-ten as 3,567.

We are all so used to thinking that the decimal representation of a number "is" that number that it rarely (if ever) occurs to us that "two" represented as 10 is just as much "two" as 2. There is no special superiority in ten as a base. The superiority of modern arithmetic lies not in the ten but in the zero. Often positional arithmetic can be equally efficient, and sometimes more efficient, with bases other than ten.

It has been said, in fact, that with the exception of nine, ten is probably the worst possible base for efficient number representation. (It is better than nine because while nine has only one divisor, ten has two.) A number like twelve, which has four divisors, would be a much more practical base, easily falling into halves, fourths, thirds and sixths. There have been many advocates, including some royal ones, of "counting by the dozen." (A thorough argument for twelve as a base instead of ten is presented in *New Numbers* by F. Emerson Andrews [Essential Books, New York, 1944].) Some mathematicians have even expressed a preference for a prime base with no divisors except the trivial ones of p and 1. Other mathematicians have still another preference. Why not a number system based on a power of two?

Two itself or its square as a base would make representation, even of relatively small numbers, too lengthy. A system based on the fourth power of two, on the other hand, would involve as we have see the addition of six new symbols. What then about a system based on the third power of two, eight? A base-eight system does not seem like a bad idea to a great many people who work with numbers. Representation in it would be almost as compact as in base-ten, and the multiplication table would be slightly smaller. Halves, fourths and eighths could be easily computed.

In spite of these arguments, it is quite unlikely that twelve, eleven, seven or eight—or any other number—will ever replace ten as the commonly used base for number representation. But the fact that they *could* serves to remind us of something that we are quite likely to forget. *A number and its symbol are not the same thing.* "Two-ness" must not of necessity be represented by 2. Whether the symbol 2 is omitted entirely in a system of number representation, as in the binary; whether a 2 in the representation of a number stands for two powers of seven, two powers of twelve, or two powers of eight instead of the usual two powers of ten; even whether some totally different symbol such as *b* is substituted for the familiar 2—the concept of the number two, the pair, remains unchanged. Two remains an interesting number.

· PROBLEMS IN BINARY · · ARITHMETIC ·

It takes a little practice to perform even the simplest operations on numbers in a system other than the decimal, but there is a pleasant feeling of satisfaction in being able to do so. Below are examples of addition, subtraction, multiplication and division as they are performed in the binary arithmetic, and then similar problems for the reader.

Addition:

$$\begin{array}{r} 100001 \\ +\ 1011 \\ \hline 101100 \end{array} \quad \text{or} \quad \begin{array}{r} 33 \\ +\ 11 \\ \hline 44 \end{array}$$

Subtraction:

$$\begin{array}{r} 11110 \\ -\ 1010 \\ \hline 10100 \end{array} \quad \text{or} \quad \begin{array}{r} 30 \\ -\ 10 \\ \hline 20 \end{array}$$

Multiplication:

$$\begin{array}{r} 1011 \\ \times\ \ 11 \\ \hline 1011 \\ 1011\ \\ \hline 100001 \end{array} \quad \text{or} \quad \begin{array}{r} 11 \\ \times\ 3 \\ \hline 33 \end{array}$$

Division:

$$\begin{array}{r} 0.010101\ldots \\ 11\,\overline{)1.000000} \\ 11 \\ \hline 100 \\ 11 \\ \hline 100 \\ 11 \\ \hline 1 \end{array} \quad \text{or} \quad \begin{array}{r} 0.333\ldots \\ 3\,\overline{)1.000} \end{array}$$

1. Add 110010 and 1111.
2. Subtract 11001 from 110111.
3. Multiply 1010 by 101.
4. Divide 1 by 101.

· ANSWERS ·

1. 1000001 (65 in the decimal system). 2. 11110 (30 in the decimal system). 3. 110010 (50 in the decimal system) 4. 0.00110011. . . (0.2 in the decimal system).

3

· THREE ·

Three is an interesting number because it is the first *typical* prime and the primes are, as a group, the most interesting of numbers.

"It would be difficult," said the mathematician G. H. Hardy, "for anyone to be more profoundly interested in anything than I am in the theory of primes."

The lure of the primes has been felt as well by many who are not professional mathematicians. For the primes are, after all, just numbers—numbers like two and three that are divisible only by themselves and one. They are the numbers from which, by multiplication, all the other numbers can be constructed, and for this reason they are often called "the building blocks of the number system." With the exception of two, they are odd, since all even numbers after two are divisible by the prime two and are hence composite. Three is thus, while not the first prime, the first typical prime.

The distinction between the two types of numbers, those that build and those that are built, came relatively late in mathematics and yet it is still an ancient one. The first definition of a prime appears in the *Elements* of Euclid (c. 300 B.C.). Much earlier, though, it was noted that some numbers are *rectilinear* (their units being capable of arrangement only in a straight line) while others are *rectangular*:

2	3	5	7	...	but	4	6	8	9	...
○	○	○	○			○○	○○○	○○	○○○	
○	○	○	○			○○	○○○	○○	○○○	
	○	○	○					○○	○○○	
		○	○					○○		
		○	○							
			○							
			○							

The rectilinear numbers, since they cannot be divided except by themselves and one, can be arranged in only one way. The rectangular numbers can be arranged in at least two ways, a straight line or a rectangle; many, like twenty-four, admit more than one rectangular arrangement.

The distinction, whether we call it prime-composite or rectilinear-rectangular, has had no practical importance until recently. But for more than two thousand years it has exerted a hold on the mind of man simply because it suggests questions that are interesting but very difficult to answer.

Most of the questions are about the primes because answering a question about primes automatically answers a question about composite numbers as well.

The first question asked about primes, and the first that was answered, was *How many prime numbers are there?* The question, in more mathematical language, is whether the set of primes is finite or infinite. The answer has great significance for the interest of these numbers. If the primes are finite they are not nearly so interesting as if they are infinite. Theoretically, we can find out anything we want to know about a finite set of numbers by sheer physical endurance. We can even count them, no matter how many there are, because at some point there is a last one. The challenge of a finite set is merely physical. With an infinite set, the challenge is mental.

For numbers the occurence of which is regular and hence predictable, it is simple to show that they continue to appear without end. The natural numbers are infinite because we can always add one to any natural number and have another. We can always add two to an even number and have another even number, add two to an odd number and have another odd number. There is no last number, no last even number, no last odd number.

With numbers like the primes, the question of how many is much more difficult to answer. For while the natural numbers string out like beads, each the same distance from its predecessor, the same from its successor, even and odd beads alternating without exception, the prime beads occur apparently without pattern in the string of numbers.

Even (O) and odd (X) numbers
OXOXOXOXOXOXOXOXOXOXOXOXOX. . . ,

but

Prime (X) and composite (O) numbers
— — XXOXOXOOOXOXOOOXOXOOOXOO. . . .

A proof that the number of primes is infinite appeared in Euclid's *Elements* almost three hundred years before the birth of Christ. It has a quality about it—a certain mathematical beauty—that even today provokes respectful envy among professional mathematicians, who cannot help asking themselves, "Would I have thought of that if it had never been thought of before?"

Euclid was an Athenian who taught for most of his life at the school in Alexandria, which he helped to found. In *The Great Mathematicians* H. W. Turnbull writes, "The picture has been handed down of a genial man of learning, modest and scrupulously fair, always ready to acknowledge the original work of others, and conspicuously kind and patient." He was a man who devoted his time to numbers, not because they are useful, but because they are interesting. When a pupil demanded to know what he would gain by proving a theorem, Euclid ordered a slave to give him a coin "since he must make a gain out of what he learns."

Euclid's proof that the number of primes is infinite is almost as straightforward as the proof that the natural numbers are infinite. It rests upon the simple fact that if we multiply together any group of prime numbers, the immediate successor of the number n that we obtain as our answer ($n + 1$ in mathematical language) will not be divisible by any of the numbers we have multiplied. It will be either another prime or a composite number that has as each of its factors a prime not in the group of primes we multiplied. This is because no number except one, which is not a prime, can possibly divide both n and $n + 1$.

When we multiply together any group of primes selected at random and add one, we can observe that a new prime does indeed result; but it is most pertinent to an understanding of Euclid's proof that the primes are infinite to observe the result when we multiply a group of consecutive primes, beginning with two and three, the first members of the set:

$$2 \times 3 = 6 \text{ and } 6 + 1 = 7, \text{ another prime,}$$
$$2 \times 3 \times 5 = 30 \text{ and } 30 + 1 = 31, \text{ another prime,}$$
$$2 \times 3 \times 5 \times 7 = 210 \text{ and } 210 + 1 = 211, \text{ another prime.}$$

If, instead of adding one to the product of "all" the primes, we subtract one, we will get a similar result:

$$2 \times 3 = 6 \text{ and } 6 - 1 = 5, \text{ another prime,}$$
$$2 \times 3 \times 5 = 30 \text{ and } 30 - 1 = 29, \text{ another prime,}$$
$$2 \times 3 \times 5 \times 7 = 210 \text{ and } 210 - 1 = 209,$$

not another prime itself but a number that has as factors the primes 11 and 19 that have not been included in our set of "all."

Euclid's proof that the number of primes is infinite is simply this. If we take the set of what we shall call "all" the primes, multiply them together, and add one to our product, we will have (as above) either another prime or a composite number with a prime factor that was not included in our set of "all." Obviously, then, we could not have had all the primes in our set. We have now generated another prime, and no matter how many primes we include in our set of "all," we can always generate still another prime in this same way.

The number of primes is, therefore, infinite.

The composite numbers are also infinite. With each additional prime we can build a composite number that we did not have before. We can, in fact, build an infinite number of infinite sets of composite numbers. An example taken from the very beginning of the natural numbers will be enough to show how the composite numbers multiply with the addition of just one prime to the set of primes. Taking two as the only prime, we have as composite numbers only the powers of two:

4, which is 2×2,

8, which is $2 \times 2 \times 2$,

16, which is $2 \times 2 \times 2 \times 2$,

. . . .

But these powers of two are infinite in number.

With the addition of three to the set of primes, we add another set of composite numbers, the powers of three:

9, which is 3×3,

27, which is $3 \times 3 \times 3$,

81, which is $3 \times 3 \times 3 \times 3$,

. . . .

These too are infinite in number. By adding three to the primes, we have also added another infinite set of numbers: each of the powers of two multiplied once by three:

12, which is $2 \times 2 \times 3$,

24, which is $2 \times 2 \times 2 \times 3$,

48, which is $2 \times 2 \times 2 \times 2 \times 3$,

. . . .

In fact—and this is easily said but difficult to grasp in all its enormousness—with the addition of three, or any prime, to the set of primes we increase the set of composite numbers by an infinite number of infinite sets. Just as we multiplied each of the powers of two by three, we can also multiply each of them in turn by each of the powers of three, of which there are an infinite number.

At this point we might as well take a deep breath and admit that there are a lot of composite numbers.

How then does the number of primes compare with the number of composite numbers?

The primes, with two and three, start out in the lead, are even at thirteen, behind at seventeen, and continue to fall farther and farther behind. They become steadily rarer while the composite numbers become steadily more numerous. There are places in the unending series of natural numbers where we have a million, a billion, a trillion, "as many as we please" composite numbers without one prime among them. These are what are called "prime deserts," and their existence can be easily proved without sending one mathematical expedition to this no man's land of number.

"As many as we please" is a favorite expression in the theory of numbers. Although it sounds like unwarranted boasting, it is not. When we say that among the natural numbers there are sequences of consecutive composite numbers "as many as we please," we mean exactly that. Let us say, for simplicity's sake, that we please to have five composite numbers occurring in succession. We first multiply together the numbers from one to six (one more than the five numbers we are after) and obtain the product 720. We know then for a certainty that the five consecutive numbers 722, 723, 724, 725 and 726 are composite.

How do we know this? We know that two divides 720, since it was one of the numbers multiplied to produce it; if it divides 720, it must also divide 722. Therefore, 722 is composite, being divisible by at least one number other than itself and one. Since three divides 720, it must also divide 723; four must divide 724; five, 725; and six, 726. We have found "as many as we please" (which in this case happens to be five) consecutive numbers that are not prime. In this particular example 721 is also composite, but generally we must assume that the immediate successor of our product may be prime, for on the basis of our proof we know of no divisors for it except itself and one.

By exactly the same process, if we please a million instead of five, we can find a place in the sequence of natural numbers where there are at least a million composite numbers between primes. Yet there is never a number beyond which all numbers are composite. In addition, even though we can prove the existence of consecutive composite numbers "as many as we please," mathematicians have not been able *to prove* that there is ever a point beyond which pairs of primes separated by only one composite number cease to occur. (Two and three, which are of course the only primes not separated by a composite number, are sometimes referred to as "the Siamese twin primes.")

As in the case of so many sets of numbers to be described in this book, the computer and its operators have turned up "twin primes" of sizes previously undreamed of, perhaps even by those who originally asked "How many?" In the first edition of this book (1955) the largest known pair of prime twins was reported as 1,000,000,009,649 and 1,000,000,009,651. In the fourth edition (1992) the largest known twins were $1{,}706{,}595 \times 2^{11{,}235} \pm 1$, discovered in 1989 by a group calling themselves "the Six of Amdahl." These were numbers of 3,389 digits compared to the 13 digit prime twins listed as "largest" in the first edition. With this edition I have decided not even to mention the largest twins since, whatever they are, they will quite likely have been exceeded within a month, or even a week. Instead, in this case, as in the case of other numbers that are constantly being hunted down, I will refer the reader to the internet for the latest discoveries.

"Almost all" numbers are composite, but there are infinitely many prime numbers.

Although it is often exceedingly difficult to determine whether a particular number is prime or composite, it is very easy to "make up" a number that we know in advance will be composite. We simply multiply a few primes together, and there we are with a composite number. We can't do anything at all similar with primes. This is because no one has ever been able to determine a form of number that is always prime.

There have been a great many attempts to find such a generating form for primes. Not one has been successful.

How then can we tell whether a number is prime?

This is one of those deceptively simple questions in which the theory of numbers abounds. The general method for testing a number for primality is implicit in the distinction between prime and composite numbers: if we can divide a number, it is not prime. We can test the primality of any number by the simple expedient of trying to divide it by each of the primes below its square root. At least one of the prime factors of a number must be equal to or smaller than the square root, since if all the prime factors were greater than the square root their product would be greater than the number itself. In the case of ninety-seven, this means trying two, three, five and seven. If ninety-seven is not divisible by any one of these four primes, it is not divisible by any number except itself and one.

There is a sort of assembly-line variation of this test of primality known as the Sieve of Eratosthenes. Eratosthenes, who lived from about 276 to about 194 B.C., is remembered particularly for an amazingly accurate measurement of the earth. His sieve seems to have been the first methodical attempt to separate the primes from the composite numbers, and all subsequent tables of primes and of prime factors have been based on extensions of it.

The compilation of such a table involves a fantastic amount of work, which is not always rewarded. One table, published in 1776 at the expense of the Austrian imperial treasury, is reported to have had such a poor sale that the paper on which it was printed was confiscated and used in cartridges in war with Turkey.

Using the Sieve of Eratosthenes, we can find all the primes under one hundred by eliminating after two every second number; after three, every third number; and so on. This leaves us with the following numbers, which are all prime:

X	X	2	3	X	5	X	7	X	X
X	11	X	13	X	X	X	17	X	19
X	X	X	23	X	X	X	X	X	29
X	31	X	X	X	X	X	37	X	X
X	41	X	43	X	X	X	47	X	X
X	X	X	53	X	X	X	X	X	59
X	61	X	X	X	X	X	67	X	X
X	71	X	73	X	X	X	X	X	79
X	X	X	83	X	X	X	X	X	89
X	X	X	X	X	X	X	97	X	X

Besides this sieve, which facilitates finding all the primes within certain limits, and the arduous method of dividing into a particular number all possible prime divisors, there is only one completely general test for primality. The theorem that states this test bears the name, not of a great mathematician, but of a young student who subsequently gave up mathematics for law. John Wilson (1741–1793) attended Cambridge University, and it was recorded by one of his professors, Edward Waring (1734–1798), that Wilson there stated what has since become known as Wilson's theorem:

> If a number n is greater than 1, then $(n-1)! + 1$ is a multiple of n if and only if n is prime.

The theorem known as Wilson's is beautifully and completely general. It can be applied as a test of primality to any number, and any number that passes the test is prime. There are more useful tests for primality than Wilson's, but none has this same quality of generality.

It is not thought that Wilson had proved his theorem. He had probably arrived at it by a little computation. The same theorem, it is now known, had already been stated but not published by Leibnitz. Later it was proved by several men whose names are also immortal in mathematics; however, it continues to bear the name of the young student who first enunciated it. Wilson, by the time his theorem was proved, was a judge; and if mathematics owes him a further debt it has not been recorded.

To test a number for primality according to Wilson's theorem, we must first compute $(n-1)!$—in words, such an expression means the product of all the numbers up to and including the immediate predecessor of the number n that is being tested. The expression $n!$ represents what is called a "factorial number," the exclamation mark being the mathematical symbol for the factorial. If the prime we wish to test is seven, the product we must first obtain is $(7-1)!$ or 6!—which is $1 \times 2 \times 3 \times 4 \times 5 \times 6$, or 720. According to Wilson's theorem, 7 is prime if, and only if, it divides evenly $(n-1)! + 1$, or 721. Since 7 does divide 721 exactly 103 times, we know that it is prime.

The trouble with Wilson's theorem is that it is more beautiful than useful. The great difficulty is not the size of the numbers involved, although they do get very large very fast, but the number of different operations that must be performed. Just take the time to compute a relative small factorial number like 26!—the number of different ways in which the letters of the alphabet can be arranged. It is pleasant to know that a number like 170,141,183,460,469,231,731,687,303,715,884,105,727 is prime if, and only if, it divides 170,141,183,460,469,231,731,687,303,715,884,105,726! + 1, but even in the theory of numbers, which is not distinguished for placing a premium on usefulness, this is not considered very useful information.

The primality of this quite long number (for it is, as it happens, prime) was determined by a completely different method. Worked out in 1876 by Edouard Lucas (1842–1891), the method, like Wilson's, tests primality without trying *any* of the possible divisors. For this reason we may discover that a number is not prime, and therefore is divisible by some number other than itself and one, and yet still not know any number that divides it.

According to Lucas, a number N of the form $2^n - 1$, where n is greater than 2, is prime if, and only if, it divides the $(n-1)$st term of a series in which the first number is 4; the second, the square of the first minus 2; the third, the square of the second minus 2—in other words, 4, 14, 194, 37634, and so on. To test the primality of 7 by this method, we must divide it into the $(n-1)$st term of the series—which, since n in the case of 7 is 3, is the second number, or 14, and find it divides evenly and is therefore prime. To test the next number of this form, which is $15 = 2^4 - 1$, we divide 15 into the third term of the series, 194, and find that it does not divide evenly and is therefore composite. The next, 31, however, divides 37,634 and so is prime.

Even Lucas's method of testing primality becomes rather unwieldy when, as in the case of $2^{127} - 1$, we must divide a number like 170,141,183,460,469,231,731,687,303,715, 884,105,727 into the 126th term of the series to determine if it is prime. For numbers of such size, Lucas worked out a shortcut. Instead of squaring each term of the series, he squared only the remainder after he had divided the number being tested into it. With this shortcut, he was able to announce at the same time he announced his new method that he had tested $2^{127} - 1$ and found it prime.

The shortcut is particularly well suited to machine calculation. In 1952 it was used in the first successful computer test of primality, which will be treated in "Six." The largest number the primality of which it established was $2^{2,281} - 1$. In the binary system that number is represented by 2281 ones.

It used to be customary to give some idea of the size of large primes by saying that they were so many times as great as something that seemed very great in itself. But the number represented by $2^{2281} - 1$ is so large that we cannot compare it even to such a large number as that of all the electrons in the universe. The square of the number of electrons (a number so large that in it each electron is replaced by a universe of electrons) is equivalent to the relatively small prime $2^{521} - 1$.

There are probably few readers who will not feel a slight thrill at the thought that it is known that for all its size this number, like 3, is divisible only by itself and 1. But just as some people can look at a mountain and feel no urge to climb it, not even feel vicariously another person's urge, many of us can see a large number and feel no curiosity whatsoever about whether it is prime or composite. Whatever it is that makes some people test large numbers for primality is probably somewhat like the impulse that makes a person embark upon the uncomfortable enterprise of climbing a mountain. As one famous mountain climber put it when asked why he wanted to climb a certain peak: "Because it is there."

It is fortunate for the theory of primes that there are those who are interested in testing the numbers themselves for primality. Much that is now known about primes in general was first suggested by extensive work on individual primes. More interesting, however, than climbing a particular mountain is finding out about mountains. Devising an efficient general test for primality is much more interesting than testing the primality of a number, no matter how large. If there is a form that invariably generates primes, that fact will be more interesting than the form itself or the primes that are generated by it. Much more interesting than the fact that a certain unimaginably large number is prime is the fact that there is no last prime—and this was proved at a time when any man alive would have been hard put to represent a very large number.

It was this—the theory of the primes as an infinite set, not the individual prime numbers—to which G. H. Hardy was referring when he said, "It would be difficult for anyone to be more profoundly interested in anything than I am in the theory of primes."

· THE POWERS OF THREE ·

If we have three weights equal to the first three powers of three (1, 3, and 9) and if we are allowed to put weight in either pan to balance our scale, we can weigh any number of pounds from 1 to 13 inclusive. In the illustration below a square □ represents the amount being weighed and a circle ○, the weight:

	Left		Right		
1			(1)		
2	(1)		(3)		
3			(3)		
4			(3)	(1)	
5	(1)	(3)	(9)		
6	(3)		(9)		
7	(3)		(9)	(1)	
8	(1)		(9)		
9			(9)		
10			(9)	(1)	
11	(1)		(9)	(3)	
12			(9)	(3)	
13			(9)	(3)	(1)

1. How many pounds can we weigh if we are allowed under the same conditions the first four powers of 3 as our weights?

2. How many if we are allowed the first five powers of 3 as weights?

3. With the knowledge of the number of pounds we can weigh with three, four and five powers of 3 respec-

tively, can you work out the general formula that will tell you how many pounds you can weigh when you are allowed *n* powers of 3 as your weights?

· ANSWERS ·

1. With four weights we will be able to weigh up to and including forty pounds. 2. With five weights we will be able to weigh up to and including 121 pounds. 3. The answer to the general question is expressed by $\frac{3^n-1}{2}$ where n is the number of powers of 3 used as weights.

· FOUR ·

Two times two is four. This is the most interesting fact about the number four, and it is very interesting indeed. Four (if we ignore the trivial 0^2 and 1^2) is the first perfect square. Four is 2^2.

There is something very solid about the symmetry of four. One of the first and most permanent number ideas was of four as the "earth number." There are still the four winds and the four elements and, of course, the four corners of the earth. Long after the world has been proved round, four carries in countless common expressions a reminder of the time when it was thought square.

The word *square* as applied to a number is a legacy from the Greeks, who looked at numbers with the eyes of geometricians. The squares to them were those numbers the units of which could be arranged in quadrilateral figures with equal sides. Arranging the units of numbers into such shapes began, according to legend, with the early Pythagoreans, who on the sand arranged pebbles into the form of people, animals or geometric forms and then assigned the number of the total to each representation. The numbers with the shape of a square are, they noted, related to other numbers in several interesting ways. Each square, for instance, is the summation of successive odd numbers; and this is the way the whole series of squares can be built up, layer by layer, from a single unit. Each square is also the product of one of the natural numbers multiplied by itself.

$$= 1 = 1 \times 1 = 1^2$$

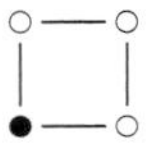

$$= 1 + 3 = 2 \times 2 = 2^2$$

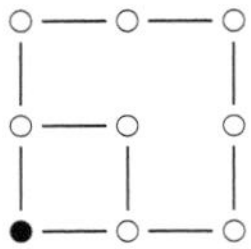

$$= 1 + 3 + 5 = 3 \times 3 = 3^2$$

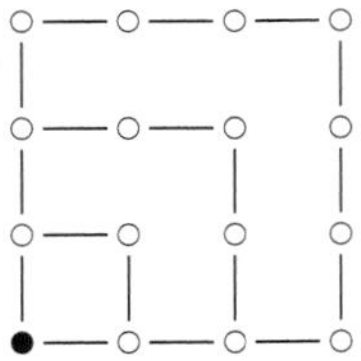

$$= 1 + 3 + 5 + 7 = 4 \times 4 = 4^2$$

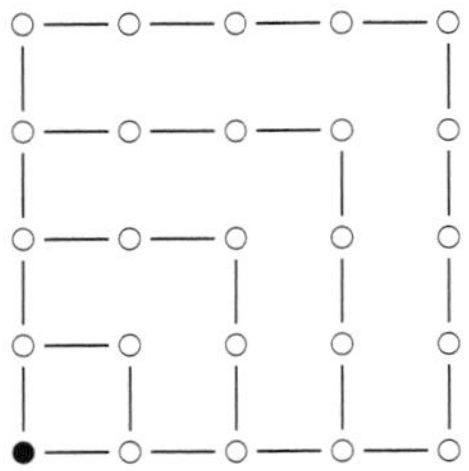

$$= 1 + 3 + 5 + 7 + 9 = 5 \times 5 = 5^2$$

. . . .

As a way of thinking of the squares, the representation of the second power of a number as n^2 has long since replaced the geometric arrangement of the units, but the name that the eye-minded Greeks gave them has persisted.

The relationships expressed above, however, are not the sort of thing that has kept four and the other squares interesting for more than two thousand years. Although fascinating to a people who were looking at numbers with fresh eyes, they are easily perceived and easily proved.

But difficulty in proving and difficulty in perceiving relationships between the squares and the other numbers are not the only criteria for mathematical interest. Let us consider a surprising relationship between the squares and the natural numbers that can be perceived merely by looking at them and is so obvious that it needs only the simplest proof.

Every number has a square. That fact does not even require a proof since it is implicit in the definition of a square as the product of a number multiplied by itself. If then every number has a square, the number of squares, like the number of numbers, is infinite. This was known to the Greeks. It is much more easily grasped than the idea that the number of primes is infinite, an idea that they also grasped and proved. Yet the fact that the squares, like the numbers themselves, are infinite suggested nothing more to them or to mathematicians after them until the time of Galileo Galilei (1564–1642).

Although the *Encyclopedia Britannica* lists Galileo as an astronomer and experimental philosopher—and this is the way we generally think of him—he was actually a professor of mathematics. The squares and the natural numbers, both unending, suggested to him a relationship that, more than two hundred years after his death, was to be basic to the development of the theory of the infinite. With this hint, the reader may be interested in seeing if he too will perceive the relationship implicit in the numbers below:

0	0^2	0
1	1^2	1
2	2^2	4
3	3^2	9
...	...	...

What Galileo saw was that with the natural numbers we can *count* the squares. The zeroth square is 0; the first, 1; the second, 4; the third, 9; and so on. The disparity between the numbers with which we are counting and the squares that are being counted becomes greater as the squares become larger—the tenth square, for example, being 100. But the important thing is that we will never run out of squares. There is a square for every natural number. The set of squares can be placed in one-to-one correspondence with the set of natural numbers in exactly the same way that back at the beginning of our understanding of number we placed two wolves in one-to-one correspondence with the wings of a bird.

There is a certain difference. Wolves and wings are finite and in our example there were exactly two of each. Numbers and squares are infinite. Yet there are obviously many more numbers than there are squares, for the squares occur less and less frequently the higher we go among the numbers. We do not have to go very high to see that this is true:

$$\boxed{0}, \boxed{1}, 2, 3, \boxed{4}, 5, 6, 7, 8, \boxed{9}, 10, 11, 12, 13, 14, 15, \boxed{16}, \ldots$$

How Galileo resolved this contradiction is explained through a character called Salviatus in his *Mathematical Discourses and Demonstrations.* Having stated what we have just noted, that the squares can be placed in one-to-one correspondence with the natural numbers, a square to every number, Salviatus comes to the conclusion:

"I see no other decision that it may admit, but to say that all Numbers are infinite; Squares are infinite; and that neither is the multitude of Squares less than all Numbers, nor this greater than that; and in conclusion, that the Attributes of Equality, Majority, and Minority have no place in Infinities, but only in terminate quantities."

This conclusion of Galileo's provides modern mathematics with one of its most important definitions. On the basis of what Galileo perceived in the relationship between the squares and all the numbers we now say:

> A set is called infinite when it can be placed in one-to-one correspondence with a part of itself.

This definition is just as true of the infinite set of squares as it is of the infinite set of natural numbers. If we divide the squares into even and odd, we find that we can place the members of the two subsets in one-to-one correspondence with the set of all the squares.

Even Squares	*Odd Squares*	*All Squares*
0	1	0
4	9	1
16	25	4
36	49	9
64	81	16
...	...	...

We will never run out of squares; neither will we run out of even or odd squares. We can rest assured—squares are inexhaustible.

Problems concerning squares are also inexhaustible. Even if as a group they were not, there would still be a satisfactory collection of individual problems that have been keeping mathematicians busy for a good many centuries and from all indications will continue to keep them busy. A case in point is the problem of the squares connected with what is undoubtedly the best known theorem in mathematics:

> The square of the hypotenuse of a right triangle is equal to the sum of the squares of the other two sides.

The Pythagorean theorem, as it is usually known, was stated and proved either by Pythagoras or by one of his followers some five hundred years before the birth of Christ. It was, like most of the Greek statements about numbers, geometrical. It posed, though, an interesting arithmetical problem. What are the solutions in whole numbers for the equation below?

$$a^2 + b^2 = c^2$$

One solution had been known for a long time. There is even a story that the Egyptians built their pyramids by marking off a rope into three, four and five units so that it fell automatically into a right triangle. This, however, is considered mathematical folklore. As mathematicians point out, there is a simpler method in Euclid I,11.

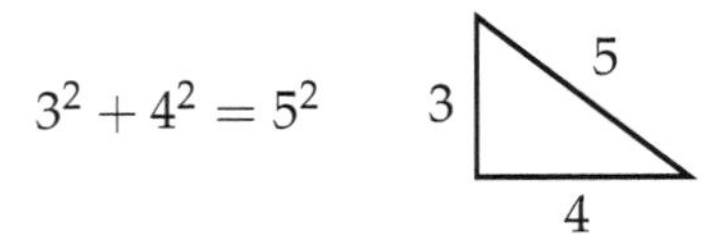

There is a way of ascertaining all possible primitive solutions to this problem. It was probably known even to the Pythagoreans, but that was by no means the end of the problem of the right triangle. Because of its relation to the squares, the Pythagorean triangle (as the right triangle with integral sides came to be known) was for centuries the basis for countless problems that, although expressed geometrically, are in reality arithmetical. Some seven centuries after Pythagoras, such problems, along with many others concerning squares and the higher powers, appeared in a little book prepared by a man known as Diophantus of Alexandria, a man whose name was to be forever linked with the squares.

Diophantus was a Greek who had an un-Greek interest in something very like algebra. Little is known about him except the problems he proposed—so little, in fact, that even the time when he lived can be estimated only in relation to the lives of other men who did or did not refer to him in their own writings. His tombstone, which proposes one last problem, tells us all that is known about his personal life:

"Here you see the tomb containing the remains of Diophantus, it is remarkable: artfully it tells the measures of his life. The sixth part of his life God granted for his youth. After a twelfth more his cheeks were bearded. After an additional seventh he kindled the light of marriage, and in the fifth year he accepted a son. Elas, a dear but unfortunate lad, half of his father he was and this was also the span a cruel fate granted it, and [his father] controlled his grief in the remaining four years of his life. By this device of numbers tells us the extent of his life."

If x is taken as the age of Diophantus at the time of his death, the problem becomes one of solving for x in the equation

$$\frac{x}{16} + \frac{x}{12} + \frac{x}{7} + 5 + \frac{x}{2} + 4 = x.^*$$

This is not the type of problem that has come to be known as a Diophantine problem. It is much too simple, there being but a single possible value for x. A more typical Diophantine problem is the ancient one of the Pythagorean triangle: to find whole number solutions for the equation $a^2 + b^2 = c^2$.

The squares, particularly in connection with this very problem, were great favorites with Diophantus. One of his problems is especially interesting because it provoked, as we shall see, a conjecture that has been the most difficult to prove in the history of the theory of numbers—at least the most famously difficult. The problem appears in Book II of Diophantus's *Arithmetic* as Problem 8: "To divide a given square number into two squares." This is the same as saying, "Given the square of the hypotenuse of a right triangle, find the squares of the other two sides"—just another of the apparently inexhaustible variations on that ancient problem.

This problem in the *Arithmetic*, and others like it, were read and struggled over for centuries before a translated version came into the hands of the man for whom they were all unknowingly destined. For Diophantus of Alexandria, who died during the third century after Christ, had the honor nearly fourteen hundred years later of introducing to numbers the man who was to become the father of modern number theory.

Pierre Fermat (1601–1665) was a busy, successful lawyer, thirty years old, when a copy of the *Arithmetic* fell into his hands. Up until that time he had apparently shown no

---

*$x = 84$.

more than a cursory interest in numbers, and he was a little old to develop a serious one. Great mathematics has most often been produced by young, sometimes even very young men. We think of poets dying young, already having achieved immortality in literature. Christopher Marlowe was twenty-nine; Shelley, thirty; Keats, twenty-six. But they died no younger than some great mathematicians. Galois was twenty when he was killed in a Paris duel; Abel died in poverty in Norway at twenty-seven. Both men left behind enough great mathematics to assure them permanent places in the history of the subject. Even mathematicians who have lived a full span must often face the fact that they did their best work when they were very young. Carl Friedrich Gauss, who was known during his lifetime (as he is known today) as the "prince of mathematicians," died at seventy-eight; but he produced his *Disquisitiones Arithmeticae,* which is usually considered his masterpiece, between his eighteenth and twenty-first years. All these mathematicians, Galois and Abel when they died, Gauss when he wrote the *Disquisitiones*, were younger than Pierre Fermat when one day he picked up the *Arithmetic* of Diophantus and got his first inkling of how very interesting the numbers are.

It has been said that Fermat was the first man to penetrate deeply into numbers. Technically never more than an amateur— his profession was the law—he is nevertheless omitted from J. L. Coolidge's *Great Amateurs in Mathematics* because, Coolidge explains, "he was so really great he should count as a professional."

For recreation Fermat, the busy lawyer, worked on the ancient problems of Diophantus. Usually these asked only for a single solution, but Fermat almost always went on—often giving methods for determining all possible solutions.

Sometimes the problems suggested to him general theorems that stated deep and previously unsuspected relationships among the numbers. As a mathematician, however, Pierre Fermat had one idiosyncrasy. Although he communicated his theorems to friends in letters or noted them down in the margins of his copy of Diophantus, he almost never stated proofs. There seems to have been no special reason that he didn't. Probably, like most mathematicians, he found what he had proved less interesting than what he was trying to prove.

In connection with Problem 8 of Book II of the *Arithmetic,* Fermat put down a characteristic note in the margin. It has been said in reference to this note that if the margin of the *Arithmetic* had been wider, the history of mathematics would have been quite different. Problem 8, as we have already stated in a slightly different form, is *To divide a given square into two squares.* Fermat was very interested in the squares, but he was also interested in the other higher powers. The problem of the squares suggested to him a much more general one, a problem that involved all the powers.

"On the other hand," he wrote in the margin beside Problem 8, "it is impossible to separate a cube into two cubes, or a biquadrate into two biquadrates, or generally any power except a square into two powers with the same exponent. I have discovered a truly marvelous proof of this which, however, the margin is not large enough to contain."

What Fermat wrote in his Diophantus is the same as saying that $a^n + b^n = c^n$ cannot be solved in positive integers when n is greater than two; that is, as in the Pythagorean theorem.

Fermat's copy of the *Arithmetic* contained many other such references to proofs that were never stated. Fermat's letters to his mathematical friends were full of more. If it is curious that Fermat never offered these friends proof of the theorems that he announced with such enthusiasm, it is more than curious that they never asked for proof. With anyone else but Fermat, the theorems would have probably been discounted by future mathematicians. Without a proof a theorem is not really mathematics. In fact, it is not really even a theorem, because a theorem in mathematics is a statement that has been *proved*. But Fermat was not only one of the most perceptive mathematicians who ever lived, but a mathematician of unimpeachable integrity. In every case except the one just given, when he said he had a proof for a theorem, a proof for it has been later (usually much later) discovered. Only this one theorem, which has long been known as Fermat's Last Theorem (even though it was neither the last nor in fact a theorem but merely a conjecture), remained unproved.

It was not from lack of effort. Almost all the great academies have at some time or other offered prizes for its proof. Almost every great mathematician since Fermat has tried his hand at it. Only Gauss refused, remarking that he himself could make a great many mathematical statements that nobody could prove or disprove.

Every so often there was a rumor in the mathematical community that one of its members had proved Fermat's "theorem." In 1988 a story appeared in the New York Times that a Japanese mathematician had accomplished the feat. As in all previous instances, however, the claim was later withdrawn.

Many special cases of the theorem had been proved. It had been definitely established that for prime values of n up to 150,000 the "theorem" held. In other words, the equation $a^n + b^n = c^n$ is not solvable in positive integers when n has any prime value from 3 up to 150,000. This served to indicate, but only to indicate, that Fermat was probably right, that for any n greater than 2 the equation is not solvable in positive integers.

Of course whether Fermat was right about the theorem was not the interesting question anymore; it was whether he was right about the proof. Was he in the seventeenth century able to prove a theorem that, in spite of concentrated effort, no mathematician in the next three centuries had been able to prove?

It was thought that Fermat's Last Theorem was true, but that Fermat was probably mistaken when he said he had a proof for it. Mathematically, it no longer seemed to matter much whether it was ever proved. It had already made its contribution, for many of the most valuable weapons of modern mathematics had been forged for what had invariably turned out to be unsuccessful assaults on Fermat's famous conjecture.

Eric Temple Bell, the mathematician and popular writer on mathematics, who firmly believed that Fermat had had a proof, devoted the last years of his life to writing a history of what he called *The Last Problem*.

"Suppose that our atomic age is to end in total disaster," he wrote in the Prospectus for the book. "What problems that our race has struggled for centuries to solve will still be open when the darkness comes down?"

Most of the "great" problems were, in his view, either too ambiguous or too broad to treat as *The Last Problem*. His nomination was a problem "that anyone with an elementary-school education can understand." His book would be "a biography" of the famous, unproved "Last Theorem" of Pierre Fermat and a biography of Fermat as well.

Bell worked on *The Last Problem* for the rest of his life, signing the contract for the almost completed book on his hospital bed. Fifteen days later on December 20, 1960, he died.

A few years later the English edition of the published book fell into the hands of a ten-year-old Cambridge schoolboy named Andrew Wiles (1953–), for whom it could have been destined. On October 25, 1994, Wiles, announced that he had proved Fermat's Last Theorem with a short but crucial step contributed by Richard L. Taylor (1962–). The proof, which eliminated an error in an earlier version, was extremely complicated, requiring some forty pages of detailed argument; but after several months of careful review it had been accepted by the mathematical community:

> The equation $a^n + b^n = c^n$ has no integer solutions when $n > 2$.

To some devotees of the ancient theory of numbers, the long delayed proof of Fermat's Last Theorem was a bit of a disappointment: "We had hoped it would have been simpler." One thing all agreed upon—if Fermat had had a proof it was not Andrew Wiles's proof!

Pierre Fermat proved many interesting things about the squares. His famous Two Square Theorem, which is cited in any discussion of mathematical beauty, states that every prime (such as five) of the form $4n + 1$ can always be represented as the sum of two squares, but that no prime (such as three) of the form $4n - 1$ can ever be so represented. Since all primes greater than two belong to one or the other of these two forms, this is a very profound statement about prime numbers.

This theorem is one of the few for which Fermat detailed his method, which he called "the method of infinite descent," although even here he did not actually give a proof. He began with the assumption that there was a prime of the form $4n + 1$ that could not be represented as the sum of two squares; proved that if there were such a prime, there must then be a smaller prime that also could not be so represented; and continued in this way until he got to five, the smallest prime of the form. Since five can be represented as the sum of two squares $(1^2 + 2^2)$, the assumption was obviously false; the theorem as stated true. (Even with this assistance from Fermat himself, the Two Square Theorem was not actually proved until almost a hundred years after his death.)

The $4n + 1$ primes, incidentally, have an interesting connection with the old problem of the right triangle. Fermat also proved a theorem that states, *A prime of the form* $4n + 1$ *is only once the hypotenuse of a right triangle; its square is twice; its cube, three times; and so on.* As an example of this theorem, in the case of five we have

$$5^2 = 3^2 + 4^2,$$
$$25^2 = 15^2 + 20^2 \text{ and also } 7^2 + 24^2,$$
$$125^2 = 75^2 + 100^2 \text{ and also } 35^2 + 120^2 \text{ and also } 44^2 + 117^2.$$

It is ironic that Pierre Fermat, who proved so many interesting things about the squares and about the other numbers, should be known for a theorem that he quite probably did not prove. In this he reminds us of Galileo Galilei, who said so many interesting things but is known for repeating stubbornly, *Eppur si muove,* which he may not have said (but probably did).

The lives of Fermat and Galileo overlapped during the years 1601 to 1642: one man in France passing busy, relatively uneventful days as a lawyer; the other in Italy, brought before the Inquisition, tried under threat of torture, recanting his deepest scientific beliefs. They led different lives; but both, like so many other mathematicians before and after them, found the squares to be very interesting numbers.

· AN OCCUPATION ·

There is nothing to keep a person occupied like trying to represent all numbers by four 4s. All four 4s must be used for every number, but various mathematical notations may also be used, as in the four examples below.

$$1 = \frac{44}{44}$$

$$2 = \frac{4 \times 4}{4 + 4}$$

$$3 = 4 - \left(\frac{4}{4}\right)^4$$

$$4 = 4 + 4 - \sqrt{4} - \sqrt{4}$$

Try now to find similar representations for 5 through 12 in the terms of four 4s.

· ANSWER ·

One possible list of answers: $5 = 4 + \left(\frac{4}{4}\right)^4$, $6 = \frac{4+4+4}{\sqrt{4}}$, $7 = 4 + 4 - \frac{4}{4}$, $8 = 4 \times 4 - 4 - 4$, $9 = 4 + 4 + \frac{4}{4}$, $10 = \frac{44-4}{4}$, $11 = \frac{44}{\sqrt{4 \times 4}}$, $12 = \frac{44+4}{4}$. There is no need to stop with 12; for it is possible, if we do not limit ourselves as to notations, to represent *all* numbers by four 4s.

· FIVE ·

One of the most interesting things about the natural numbers is that although nothing about them changes, they retain the ability to surprise us. A case in point is that of the pentagonal numbers—those that, as their name implies, can be arranged in the shape of a five-sided polygon.

The Pythagoreans had a special fondness for the shape of five, for within the regular pentagon they constructed the "triple-interwoven triangle"—the five-pointed star that was the symbol of recognition in their order. Nevertheless, to them the pentagonal numbers were just one group in an infinitude of so-called polygonal numbers that they found very interesting. These numbers began with three as the triangle, four as the square, five as the pentagon and continued without end through the natural numbers. For the Greeks observed the obvious though essentially farfetched relationship that "every number from three on has as many angles as it has units."

They then observed that they could add to each polygon a row of units and have another larger polygon of the same number of sides. Because in this row by row construction, one was the point from which construction began,

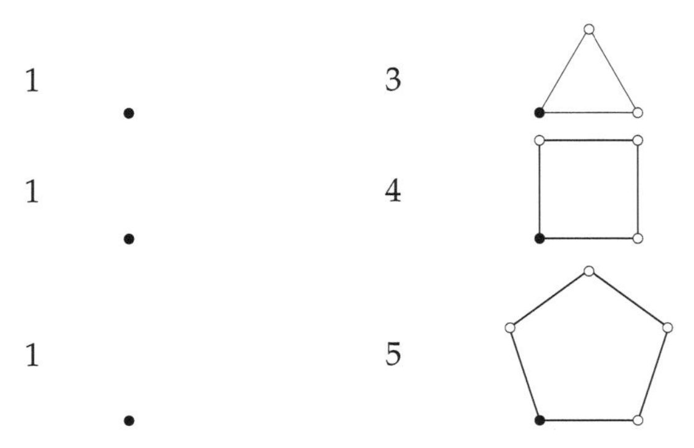

one was considered the first polygon in each group. In the case of five, for example, successive pentagons were formed from a point as follows:

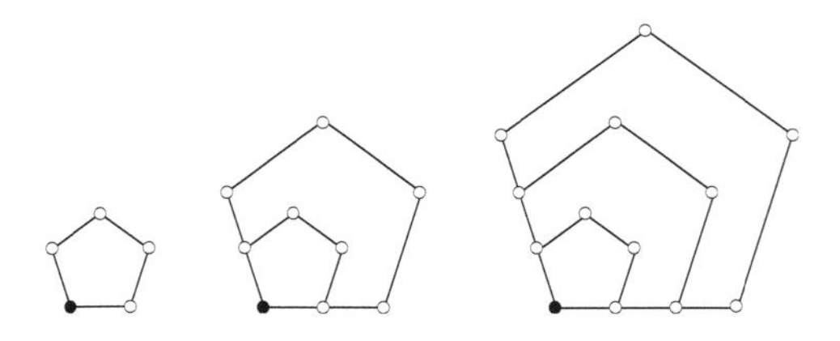

Five was thus the archetype of the pentagonal numbers, having as many units as a pentagon has angles, but one was the first. The first twelve pentagonal numbers appear below:

$$1, 5, 12, 22, 35, 51, 70, 92, 117, 145, 176, 210, \ldots .$$

Since a standard test of mathematical aptitude is the ability to see the basis for selection in a series of numbers like the one above, the reader might like to try to continue the series by adding the appropriate next number. The next number, as it happens, is the thirteenth pentagonal number. There is more than one way of arriving at its value.

The first way is by summing one and every third number up to and including the thirteenth. As we saw in "Four," the squares are successive summations of every second number after one. The pentagonals, we now see, are the successive summations of every third number after one (the hexagonals of every fourth, and so on).

$$
\begin{aligned}
\text{Pentagonal } \#1 &= 1 = 1, \\
\#2 &= 1 + 4 = 5, \\
\#3 &= 1 + 4 + 7 = 12, \\
\#4 &= 1 + 4 + 7 + 10 = 22, \\
&\vdots \\
\#12 &= 1 + 4 + 7 + 10 + 13 + 16 + 19 + 22 + \\
&\quad 25 + 28 + 31 + 34 = 210.
\end{aligned}
$$

Thus to obtain the thirteenth pentagonal number, we add 37 (34 + 3) to the twelfth pentagonal number, which gives us 247.

The second way of arriving at the thirteenth pentagonal number is more direct, but it involves a knowledge of the general formula for determining any polygonal number. In the language of mathematics, any polygonal number is referred to as the nth r-agonal number. In our case we are after the 13th 5-agonal number. In the formula below the letter n has the value of 13, the letter r of 5:

$$
p_n^r = \frac{n}{2}\left[2 + (n-1)(r-2)\right] \text{ or } n + (r-2)n\frac{(n-1)}{2}
$$

or

$$
\text{13th 5-agonal number} = \frac{13}{2} \times 38 \text{ or } 13 + (3 \times 78) = 247.
$$

Like many after them who have looked at numbers for the first time, the Greeks found the formation and interrelationships of the polygonal numbers very interesting. More sophisticated mathematicians, once they have the above formula with which they can obtain any polygonal number of any rank, are inclined to dismiss them as the kind of thing that only amateurs would find interesting. But no one can dismiss as uninteresting numbers that Pierre Fermat found interesting. Fermat looked at the natural numbers with something of the fresh interest of the Greeks, but in the case of the polygonal numbers he went below the surface relationships that had intrigued the Greeks and so discovered a relationship between the polygonals and all the numbers of which they had never dreamed.

"Every number," wrote Fermat, again in the margin of his copy of Diophantus, "is either triangular or the sum of two or three triangular numbers; square or the sum of two, three, or four squares; pentagonal or the sum of two, three, four, or five pentagonal numbers; and so on."

The beauty of this theorem lies in the *every number* and the *and so on*. It is completely general. It says something about all numbers and about all polygonal numbers, and what it says is not at all obvious. J. V. Uspensky and M. A. Heaslet, who in their *Elementary Number Theory* dismiss the Greek interest in the polygonal numbers as trivial, say of Fermat's theorem, respectfully, "This is a truly deep property of numbers."

That every number can be expressed as the sum of five or fewer pentagonal numbers is an unexpected relationship between these numbers of the shape of five and all the natural numbers. It lacks, however, that quality of surprise that we find in an even later discovery.

The discoverer in this case was no amateur in mathematics, even in the grand sense of Pierre Fermat. Leonhard Euler (1707–1783) was one of the most completely professional mathematicians who ever lived and, without a rival, the most prolific. During his seventy-six years there was scarcely an aspect of mathematics that he did not leave more systematized than he had found it, this in spite of the fact that during the last seventeen years of his life he was almost totally blind.

It was one of the rare things about Euler that he did what needed to be done in mathematics—whether it was making a great original contribution or merely picking up pins to keep things straight. One thing that was needed at the time of Euler—and the thing we are especially interested in—rose out of the study of partitions. It was there, where no one would have expected to find the pentagonal numbers playing an important role, that Euler found them.

In the theory of partitions, we are concerned with the number of ways in which a number can be represented as the sum of its parts. The partitions can be restricted to particular parts (such as odd parts or distinct parts), but most generally they are all parts without restriction. How many different ways, for instance, can the number five be represented as the sum of the numbers one, two, three, four and five?

$$5 = 5,$$
$$5 = 4 + 1,$$
$$5 = 3 + 2,$$
$$5 = 3 + 1 + 1,$$
$$5 = 2 + 2 + 1,$$
$$5 = 2 + 1 + 1 + 1,$$
$$5 = 1 + 1 + 1 + 1 + 1.$$

The number of unrestricted partitions for five is seven, or $p(5) = 7$.

The general problem in the theory of partitions is to determine the number of partitions possible for each of the natural numbers. This is no simple task, for the number of partitions bears no fixed relationship to the number being partitioned. One can be partitioned in only one way, two in two ways, three in three ways, but after three the relationship is no longer one-to-one. The number of partitions for four is five; for five, as we have seen, seven. The reader may be interested in hazarding a guess as to the number of ways in which six can be partitioned. Unless he actually computes them, he is almost certain to come up with the wrong answer.[1]

If there is no apparent relationship between a number and the number of its partitions, how then can we determine the partitions of a particular number without actually computing them? What would be useful would be a combination of numbers that by multiplication or division, or both, would automatically produce the answers we are after. *Producing answers indefinitely.* For we want the number of partitions for each and every one of an infinitude of numbers.

It was just such a "generating function" that Euler contributed to the theory of partitions. In discovering it, he discovered also a very surprising relationship between the pentagonal numbers and the unrestricted partitions of all the natural numbers.

Euler's generating function for $p(n)$ is the reciprocal of a power series:

---

[1] $p(6) = 11$.

$$\frac{1}{(1-x)(1-x^2)(1-x^3)(1-x^4)(1-x^5)\ldots}.$$

The expression indicates, as does any fractional expression, that a division is to be performed. There will, however, be something very odd about this division. We will have to start our division without ever having completed our divisor. We will be instructed by the three dots after $(1-x^5)$ that we are to continue multiplying in the same fashion. Each time we are to increase the power of x by 1 so that we will be multiplying $(1-x^5)$ by $(1-x^6)$ by $(1-x^7)$ and so on and on. The terms being multiplied will never end, nor will the product that we obtain with them. More and more of it will be finally established the longer we multiply, but all of it—never. It must be what is known as an infinite product.

When therefore we divide this infinite product into one, we will obtain as our answer an infinite quotient. This is as it should be, for the generating function for $p(n)$ must by definition never stop generating. It has to produce for us the number of partitions for every one of the infinitude of natural numbers.

All in all, this generating function for $p(n)$ is an odd sort of thing to anyone used to multiplication and division with finite numbers. The multiplication that gives us the infinite product looks odd enough, but the division that gives us the infinite quotient looks odder yet.

We begin, with deceptive ease, by multiplying $(1-x)$ by $(1-x^2)$. The early part of the multiplication is printed in full below so that the reader can see how in a short time the first few terms of our answer no longer change with continued multiplication. They are, in other words, established; and we can use them to begin our division of 1:

$$
\begin{array}{l}
1-x\\
1-x^2\\
\hline
1-x-x^2+x^3\\
1-x^3\\
\hline
1-x-x^2+x^3\\
\qquad\qquad -x^3+x^4+x^5-x^6\\
\hline
1-x-x^2 \qquad +x^4+x^5-x^6\\
1-x^4\\
\hline
1-x-x^2 \qquad +x^4+x^5-x^6\\
\qquad\qquad\quad -x^4+x^5+x^6 \qquad -x^8-x^9+x^{10}\\
\hline
1-x-x^2 \qquad\qquad +2x^5 \qquad\qquad -x^8-x^9+x^{10}\\
1-x^5\\
\hline
1-x-x^2 \qquad\qquad +2x^5 \qquad\qquad -x^8-x^9+x^{10}\\
\qquad\qquad\qquad\quad -x^5+x^6+x^7 \qquad\qquad -2x^{10}\ldots\\
\hline
1-x-x^2 \qquad\qquad +x^5+x^6+x^7-x^8-x^9-x^{10}\ldots\\
1-x^6\\
\hline
1-x-x^2 \qquad\qquad +x^5+x^6+x^7-x^8-x^9-x^{10}\ldots\\
\qquad\qquad\qquad\qquad\quad -x^6+x^7+x^8 \qquad\qquad -x^{11}-x^{12}\\
\hline
1-x-x^2 \qquad\qquad +x^5 \qquad +2x^7 \qquad -x^9-x^{10}-x^{11}-x^{12}\ldots\\
1-x^7\\
\hline
1-x-x^2 \qquad\qquad +x^5 \qquad +2x^7 \qquad -x^9-x^{10}-x^{11}-x^{12}\ldots\\
\qquad\qquad\qquad\qquad\qquad\quad -x^7+x^8+x^9 \qquad\qquad -x^{12}\ldots\\
\hline
1-x-x^2 \qquad\qquad +x^5 \qquad +x^7+x^8 \qquad -x^{10}-x^{11}-2x^{12}\ldots\\
1-x^8\\
\hline
\end{array}
$$

...

The first of these established terms are

$$1 - x^1 - x^2 + x^5 + x^7 - x^{12} - x^{15} + x^{22} + x^{26} \dots .$$

By substituting a 1 for each of these powers of x that has remained in the product and substituting a 0 for each that has dropped out, we get an odd looking representation for the beginning of our divisor:

$$1 - 1 - 1 + 0 + 0 + 1 + 0 + 1 + 0 + 0 + 0 + 0 - 1 + 0 \dots .$$

We are now ready to divide one, or unity:

$$\begin{array}{r|l}
 & 1+1+2+3+5+7+11\dots \\
\hline
1-1-1+0+0+1+0\dots & 1+0+0+0+0+0+\ 0\dots \\
 & 1-1-1+0+0+1+\ 0\dots \\
\cline{2-2}
 & \quad +1+1+0+0-1+\ 0\dots \\
 & \quad 1-1-1+0+0+\ 1\dots \\
\cline{2-2}
 & \qquad +2+1+0-1-\ 1\dots \\
 & \qquad 2-2-2+0+\ 0\dots \\
\cline{2-2}
 & \qquad\quad +3+2-1-\ 1\dots \\
 & \qquad\quad 3-3-3+\ 0\dots \\
\cline{2-2}
 & \qquad\qquad +5+2-\ 1\dots \\
 & \qquad\qquad 5-5-\ 5\dots \\
\cline{2-2}
 & \qquad\qquad\quad +7+\ 4\dots \\
 & \qquad\qquad\quad 7-\ 7\dots \\
\cline{2-2}
 & \qquad\qquad\qquad +11\dots .
\end{array}$$

From the fragment of the division that we have reproduced, the reader will observe that the numbers appearing in our answer seem strangely familiar. They are indeed familiar, for they are—in order—the number of unrestricted partitions p for each of the first few natural numbers:

$$
\begin{aligned}
p(0) &= 1, \\
p(1) &= 1, \\
p(2) &= 2, \\
p(3) &= 3, \\
p(4) &= 5, \\
p(5) &= 7, \\
p(6) &= 11, \\
&\ldots .
\end{aligned}
$$

Continuing the division of one in this manner will go right on producing the successive values of $p(n)$. It will even turn up the fact that the number of unrestricted partitions for a relatively small number like 200 is 3, 972, 999, 029, 388.

In arithmetic like this we have no reason to expect to find our old friends—those trivial, if amusing, numbers that can be arranged into the shape of five. Yet if we will examine the first few terms of the infinite product that have been definitely established, we will find

$$1 - x^1 - x^2 + x^5 + x^7 - x^{12} - x^{15} + x^{22} + x^{26} \ldots .$$

Now let us do a little arithmetic of our own. The formula for a pentagonal number is

$$P_n^5 = \frac{3n^2 - n}{2}.$$

Up until now we have been considering only the pentagonal numbers produced by this formula when the value of n is 0 or one of the positive integers:

$$\begin{aligned} 1 &\text{ when } n = +1,\\ 5 &\text{ when } n = +2,\\ 12 &\text{ when } n = +3,\\ 22 &\text{ when } n = +4,\\ &\ldots . \end{aligned}$$

But the same formula also yields pentagonal numbers for negative values of n:

$$\begin{aligned} 2 &\text{ when } n = -1,\\ 7 &\text{ when } n = -2,\\ 15 &\text{ when } n = -3,\\ 26 &\text{ when } n = -4,\\ &\ldots . \end{aligned}$$

If we now re-examine the first few established terms of our infinite product, we find that *the only* x *that remain are those the exponents of which are the pentagonal numbers produced by the formula above for both negative and positive values of* n.

This is not a relationship that Euler or anyone else suspected. It is curious and not exactly clear why the pentagonal numbers should make this appearance in the generating function for $p(n)$. But it is a discovery that the Greeks would have appreciated. For if they were distracted sometimes by the more trivial and obvious aspects of number composition, they were nevertheless the first people to realize that the numbers as numbers are fascinatingly complex in their relationships—and full of surprises.

· ANOTHER SURPRISE ·

It is very satisfying to discover for oneself interesting relationships among the numbers even though the same relationships have already been discovered by somebody else. If the reader will square the beginning of the infinite product that we obtained in this chapter, he will find nothing interesting about the result; but if he will cube it, he will discover a surprising and interesting pattern. The first to discover this pattern was C. G. J. Jacobi (1804–1851), a very great mathematician.

$$1 - x - x^2 + x^5 + x^7 - x^{12} - x^{15} \text{ multiplied by}$$
$$1 - x - x^2 + x^5 + x^7 - x^{12} - x^{15} \text{ multiplied by}$$
$$1 - x - x^2 + x^5 + x^7 - x^{12} - x^{15} \text{ yields?}$$

· ANSWER ·

$1 - 3x + 5x^3 - 7x^6 + 9x^{10} \ldots$, the exponents that remain in this case are the triangular numbers rather than the pentagonal numbers and the coefficients are alternately the positive and negative odd numbers.

· SIX ·

Six is the first "perfect number."

The Greeks called it perfect because it is the sum of all its divisors except itself. These are one, two and three: $6 = 1 + 2 + 3$.

The Romans attributed the number six to the goddess of love, for it is made by the union of the sexes: from three, which is masculine since it is odd, and from two, which is feminine since it is even. The ancient Hebrews explained that God chose to create the world in six days instead of one because six is the more perfect number.

Numbers that are mathematically "perfect" in the ways of six—the sum of all their divisors except themselves—have been interesting to mathematicians, and to others, since the time of the Greeks. But starting out with six, mathematicians in more than two thousand years had turned up only eleven more that met the strict requirements for numerical perfection.

Then, at the beginning of 1952, as the twentieth century settled into its second half, a University of California professor, Raphael M. Robinson (1911–1995), using the computer at what was then the Institute for Numerical Analysis on the Los Angeles campus of the University, turned up the first new perfect number in seventy-five years and, in the next few months, four more that brought the total of known perfect numbers to seventeen.

Robinson's discovery did not attract the attention of the press. Perfect numbers were not useful in the construction of bombs. In fact, perfect numbers were not useful for anything at that time. But they were interesting to mathematicians (they had interested Gauss), and their story is an interesting one. It begins, like most stories in mathematics, with the Greeks, who, having noticed the fact that six $(1+2+3)$ and twenty-eight $(1+2+4+7+14)$ are both the sum of all their divisors except themselves, wondered how many numbers there were that were like them. The basic similarity of six and twenty-eight is apparent when both are represented algebraically. They are of the form $2^{n-1}(2^n-1)$:

$$6 = 2^1(2^2-1), \text{ or } 2 \times 3,$$
$$28 = 2^2(2^3-1), \text{ or } 4 \times 7.$$

Euclid, more than two thousand years ago, had proved that all numbers of this form are perfect when 2^n-1 is divisible only by itself and one or, in other words, prime. For 2^n-1 to be prime, n must also be prime. In the case of 6 we see from the above that the prime essential to its formation is 3 or 2^2-1; in the case of 28, 7 or 2^3-1. Euclid, however, did not prove that all perfect numbers are of this form, and he left for future mathematicians a question:

How many perfect numbers are there?

In the following centuries, the numbers seemed to gain more ethical than mathematical significance. L. E. Dickson

(1874–1954) in his history of the theory of numbers reports that in the first century A.D. numbers were separated into *abundant* (those like twelve whose divisors total more than themselves), *deficient* (those like eight whose divisors total less), and *perfect;* and the moral implications of the three types were carefully analyzed. In the eighth century it was pointed out that the second origin of the human race was made from the deficient number eight, since in Noah's Ark there were eight human animals from whom the entire human race sprang, this second origin being thus more imperfect than the first, which was made according to the perfect number six. In the twelfth century the study of perfect numbers was recommended in a program for "Healing of Souls."

Nobody, however, answered Euclid's question.

Nobody, in fact, seemed much concerned with the mathematics of the subject. The first four perfect numbers—6, 28, 496 and 8,128—had been known as early as the first century. The basic theorem concerning perfect numbers had been enunciated by Euclid three hundred years before the birth of Christ. Yet it was not until fourteen centuries later, in spite of all the speculation on the subject in general, that the fifth perfect number was correctly stated as 33,550,336.

Looking smugly back from the age of computers, we may forget that the discovery of perfect numbers, even those much smaller than the largest ones known today, has always involved a considerable amount of computation. Let's take this fifth perfect number as a case in point. From it the reader can get some idea of just how much of the third "R" was behind the announcement that it is perfect.

To prove that $2^{13-1}(2^{13}-1)$, the representation of 33,550, 336 according to Euclid's formula, is perfect, we must prove that $2^{13}-1$ is prime. First we must compute the number represented by $2^{13} - 1$. The multiplication of thirteen 2s gives us the product 8,192, which less 1 gives us $2^{13} - 1$ as 8,191.

To ascertain whether 8,191 is prime, we must try to divide it by all the primes below its square root, which falls between 90 and 91. There are twenty-four primes below 90. Only after we have verified that none of these divides 8,191 can we say that it is prime. For this job we need an accurate listing of the primes and accurate reckoning at every step. When we recall that reckoning up until this time was done without a practical system of arithmetic notation, we do not marvel that it was so long until the fifth perfect number was accurately stated. For *after* we have ascertained that 8,191 is indeed prime, we must multiply it by 4,096 (or 2^{12}) to obtain 33,550,336, the fifth perfect number.

The reader with time to spare may be interested in trying his skill on $2^{17-1}(2^{17} - 1)$, which is the next possible perfect number of Euclid's form.

Because perfect numbers are, after the fourth, so large and offer so many opportunities for error in computation, a great many imperfect numbers were announced at various times as perfect.

There was a tendency also to guess at the unknown numbers from the known. On the basis of the first four perfect numbers—6, 28, 496 and 8,128—two guesses were widely accepted. One was that the perfect numbers ended alternately in 6 and 8. As it happens, they do end in 6 or 8 but not alternately or in any discernible pattern. That hypothesis went by the board with the announcement of the sixth perfect number—8,589,869,056, which ends in 6 when it "should" end in 8. The other guess was that perfect numbers appeared in regular fashion throughout the numbers, one (6) in the units, one (28) in the tens, one (496) in the hundreds, one (8,128) in the thousands. The discovery of the fifth perfect number, which left the ten thousands and the hundred thousands without their allotted perfect numbers, disproved this hypothesis.

In the perfect number business, though, anybody's guess appears to have been as good as anybody else's. The fact that a guess was wrong did not write it off the records. The particular primes of the form $2^n - 1$ that are necessary for the formation of perfect numbers of Euclid's form bear for all time the name of a man who guessed wrong.

Marin Mersenne (1588–1648) was a friar whose best claim to mathematical importance lay in the fact that he was a favorite correspondent of both Fermat and Descartes. It was in 1644 that he established another claim to mathematical importance and linked his name forever with the perfect numbers. Since with the fifth perfect number even the necessary primes were enormous, it had become necessary to describe all perfect numbers by the prime values of n in the expression $2^n - 1$. The five known perfect numbers were by this system designated as 2 (for the exponent in $2^2 - 1$, the prime necessary for the formation of 6), 3, 5, 7 and 13. Mersenne now announced that there were only six more such prime exponents up to and including 257. He listed them as 17, 19—which in his time had already been shown to be prime by Pietro Cataldi (1552–1626)—31, 67, 127 and 257. This last and largest number he announced as prime ($2^{257} - 1$) is

$$231584, 178474, 632390, 847141, 970017, 375815, 706539,$$
$$969331, 281128, 078915, 168015, 826259, 279871.$$

It was obvious to other mathematicians that Father Mersenne could not have tested for primality all the numbers that he had announced as prime. But neither could they. One contemporary suggested hopefully that the basis of Mersenne's assertion was doubtless to be found in his stupendous genius, which perhaps recognized more truths than he could demonstrate.

At the time Mersenne announced his primes—primes far beyond those he listed are still designated by his name—the only method of testing the primality of a number was the one previously mentioned of actually dividing into it all the primes smaller than its square root. This was a method so time consuming that for some Mersenne numbers even the computers of the distant future would not be able to achieve results with it. But by this laborious method mathematicians did test or re-test the primes announced by Mersenne for the sixth, seventh and eighth perfect numbers. The eighth $(2^{31} - 1)$ was tested and found prime by Euler, busy as usual doing what needed to be done in mathematics.

A mathematical writer commented that the perfect number formed from Euler's prime would quite probably be the last to be discovered: "For, as [the perfect numbers] are merely curious without being useful, it is not likely that any person will attempt to find one beyond it." Little did he reckon with the curiosity of mathematicians when the question is whether a particular kind of prime number is finite or infinite.

It was Euler who made the most important contribution since Euclid to the question of the perfect numbers. Euclid had proved that any number of the form $2^{n-1}(2^n - 1)$ is perfect when $2^n - 1$ is prime, but he did not prove that all perfect numbers are of this form. He proved that all *even* perfect numbers are. There are, as far as is known, no odd perfect numbers; but it has never been *proved* that there are none.

Euler's perfect number remained the largest known for more than a hundred years. Then in 1876 Edouard Lucas (1842–1891) worked out the method that we have already described in "Three" by which a possible prime can be tested without factoring. At the same time he announced that he had tested $2^{127} - 1$ and found it prime. Although in 1891 in his *Theorie des Nombres* he changed his mind and listed the number as "undecided," after further verification in 1913 it was accepted as the largest known Mersenne prime.

Even with the help of Lucas's much more efficient method of testing, mathematicians were not able to finish testing all the Mersenne numbers until the following century. The last, $2^{257} - 1$, required over a year of work on a standard electric calculating machine and then another year for checking the result. It is not prime. Since it was the largest guess Mersenne had made, the final score on the mathematician-friar could at last be reckoned. In addition to the five perfect numbers already known at the time he made his famous announcement, he had listed four more correctly (17 and 19—both already established as prime—and 31 and 127) and two incorrectly (67 and 257); and he had omitted three that he should have included since they are below 257 (61, 89 and 107).

As the twentieth century went into its second half, there were twelve known perfect numbers, the largest being $2^{126}(2^{127} - 1)$, which had been discovered seventy-five years previously by Lucas. There had been one venture in 1951 with the newly invented computer, but it had merely verified M_{127} about which Lucas had been a bit doubtful.

Euclid's question was still unanswered.

The machine that in 1952 broke the barrier in the Mersenne numbers was the National Bureau of Standards

Western Automatic Computer, known as the SWAC. It was at that time one of the fastest computers in existence.[1] It could add two ten-digit figures in sixty-four microseconds. Since a microsecond is one-millionth of a second, this meant that the SWAC could do an addition of that type 156,000 times as fast as a human being could—if a human being could do it in ten seconds. These figures are not impressive today, but in 1952 they were very impressive. The SWAC extended man's ability to compute just as the then new Palomar telescope, which was its neighbor in southern California, extended his ability to see.

But the SWAC was no mathematician. Except for its speed and accuracy it was inferior to any human being who knew how to add, subtract, divide and multiply efficiently. For it could not compute anything it had not been told how to compute.

Robinson, in Berkeley, had never seen the machine in Los Angeles, but he set out to program the SWAC to test Mersenne numbers by using only the manual.

The job was to break down the Lucas method of testing primality into a program of the thirteen kinds of commands to which the SWAC responded. The job was complicated by the fact that while the machine was built to handle numbers of 36 bits, the numbers involved ran to 2,300 bits. The total memory of the machine was only 256 words, each word consisting of 36 bits and a sign, so a number of approximately 2,300 bits required 64 words. But for the testing by Lucas's method the number would also have to be squared. Thus one number could use up half the memory of the machine. It was, Robinson found, very much like trying to explain to a human being how to multiply hundred-

[1]The Bureau of Standards had two computers, the SEAC on the east coast and the SWAC on the west.

digit numbers on a desk calculator built to handle ten-digit numbers.

The program had to be written entirely in machine language. One hundred and eighty-four separate commands were necessary to tell the SWAC how to test a possible prime by the Lucas method. The same program of commands, however, could be used for testing any number of the Mersenne type from $2^3 - 1$ to $2^{2297} - 1$. The latter was the largest that could be handled.

There was still more to be done before the machine could "solve" the problem. The commands had to be coded. This was done by using the letters and signs of the standard typewriter keyboard, the letter "a," for example, being the code letter for the command to add. Coded, the commands were then transferred to a heavy paper tape so that they became merely an arrangement of perforations that could be recognized by the machine either as an electric impulse (a hole punched in the tape) or as the lack of an impulse (no hole).

Such simplicity of language was the main factor in the SWAC's then amazing computing speed. Even the enormous numbers it worked with were expressed wholly as 1s (impulses) or 0s (no impulses). The SWAC, instead of using the decimal system for its computations, used the binary system that we described in "Two."

D. H. Lehmer (1905–1991), the Director of the SWAC, and his staff had not been unduly excited when Robinson sent his program down to Los Angeles. After all, their own programs, in spite of their familiarity with the machine and its manual, almost never ran without error the first time. They put Robinson's program in a drawer for when they had the time to show him it would not run. But Robinson kept insisting that they run his program. Finally, after a fortnight, muttering that Robinson would find out programming a computer from the manual was not easy, they took his work out of the drawer where it had languished.

On the evening of January 30, 1952, the program of commands, coded and punched on a twenty-four-foot tape was placed in the machine. In comparison to the seconds it would take the SWAC to obey all the commands on the tape, the insertion itself took an extremely long time—several minutes. All that was then necessary to test the primality of any Mersenne number was to insert the exponent of the new number as it was to be tested. The machine could do the rest, even to typing out the result—continuous zeros if the number was a prime, a number written out to the base-sixteen if it was not.

The proof of primality would be in a string of continuous zeros because by the Lucas test (described in "Three") a number is prime only if it leaves no remainder when divided into a certain term in a certain series. The version of the Lucas test that was used by Robinson was an improvement on the test by Lehmer, the director of the Institute.

The human operator of the SWAC, sitting at a desk in front of the large machine, inserted the first number to be tested. He typed it out backwards, not in the binary system, which would have made his job too lengthy, but in the base-sixteen so that the machine itself could transpose it into the binary. He then pressed a button on the panel of his desk, and the machine, following the one hundred and eighty-four instructions it had received, began the test for primality of the first number.

The first number chosen to be tested was $2^{257} - 1$, the largest of the eleven numbers announced as prime by Mersenne. Twenty years before the SWAC test, it had been tested and found not prime by Lehmer and his wife, Emma

Trotsky Lehmer (1906). It had taken them two hours a day for a year to make the test using a handcranked electric calculating machine that made so much noise the neighbors complained if they worked at night. This evening both the Lehmers were in the room to see the machine, in a fraction of a second, arrive at the answer that had taken them an arduous seven hundred and some hours: $2^{257} - 1$ is not prime.

The SWAC then continued on a list of larger possible primes. Mersenne had said, four hundred years before, that to tell if a given number of fifteen or twenty digits is prime, all time would not suffice for the test; but he had not foreseen a shortcut like the Lucas-Lehmer method or a machine like the SWAC. One by one, by that method, the SWAC tested forty-two numbers, the smallest having more than eighty digits. Not one proved prime.

It was not until ten o'clock that evening that the long awaited string of zeros came out of the machine. The number just tested, briefly expressed as $2^{521} - 1$, was the first new Mersenne prime discovered in seventy-five years. The new perfect number that could be formed from it—$2^{520}(2^{521} - 1)$—was only the thirteenth perfect number to be discovered in almost twice that many centuries.

The fact that Robinson's program ran successfully on its first trial created something of a sensation. "That the code was without error was (and still is) a remarkable feat," wrote John Todd and Magnus R. Hestenes in their history of the Institute for Numerical Analysis.

For a period of approximately two hours on the night of January 30, 1952, $2^{521} - 1$ had the distinction of being the largest known prime number as well as the largest known Mersenne prime. Then shortly before midnight the string of zeros announcing another, larger prime, $2^{607} - 1$, came up. Over the next several months Robinson's program produced a total of five previously unknown Mersenne primes. The testing of the thirteenth Mersenne prime had taken the SWAC approximately one minute—the equivalent of a year's full-time work for a man. The seventeenth and last took the machine an hour. It would have taken a human being a lifetime.

Some thirty years later Robinson ran his program on one of the first IBM PC's, which turned out to be twice as fast as the SWAC.

In the last fifty odd years the number of known Mersenne primes (and hence the number of known perfect numbers) has almost tripled. At the time (1992) that I was working on the fourth edition of *From Zero to Infinity*, there were thirty-one, the most recently discovered Mersenne prime having been found six years earlier. But ever improving technology and the importance of enormous primes in cryptology have resulted in what might be called a great leap forward. Since 1996, when George Woltman organized The Great Internet Mersenne Prime Search—GIMPS, as it is called—has had what it describes as "a virtual lock" on the largest known prime number:

"This is because its excellent free software is easy to install and maintain, requiring little of the user other than to watch and see if they find the next big one!" GIMPS announces on its website, and adds, "Tens of thousands of users have replaced the ubiquitous inane screen savers with this much more productive use of their computers' idle time."

As I type this, pulling up the official web site[2] of GIMPS, I learn that the largest known Mersenne prime—at this

[2]http://primes.utm.edu/largest.html.

moment—is $2^{25,964,951} - 1$, the 42nd known Mersenne prime, a number of 7,816,230 digits, discovered on February 26, 2005 by a participant in GIMPS, a German ophthalmologist.

Yet, by a proof as old as Euclid, we know that the perfect number that results from this prime, unbelievably enormous though it is, is the sum of all its divisors except itself—just as surely as we know that $6 = 1 + 2 + 3$ and $28 = 1 + 2 + 4 + 7 + 14$.

On the GIMPS website Gauss is quoted as writing in the *Disquisitiones*:

"The problem of distinguishing prime numbers from composite numbers and of resolving the latter into their prime factors is known to be one of the most important and useful in arithmetic. It has engaged the industry and wisdom of ancient and modern geometers to such an extent that it would be superfluous to discuss [it] at length... . Further the dignity of the science itself seems to require that every possible means be explored for the solution of a problem so elegant and celebrated."

But how many perfect numbers are there? Is their number finite or infinite?

Euclid's question is still unanswered.

· OLD FAVORITES ·

Not quite so old as perfect numbers, but quite old, are the *amicable numbers*. These are pairs of numbers such that each is the sum of the divisors, including one, of the other. Today many pairs of these numbers are known. Euler published at one time a list of 64, two of which turned out to be false. But the ancients knew only one, a pair of numbers that they considered the symbol of perfect harmony. One member of the pair is 220, and the reader may like to see if he can determine the other.

· ANSWER ·

The divisors of 220 are 1, 2, 4, 5, 10, 11, 20, 22, 44, 55 and 110; and these add up to 284, the divisors of which add up to 220.

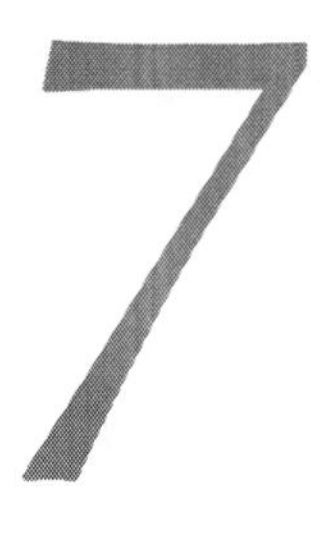

· SEVEN ·

The number seven has been held in esteem since antiquity as being unique among the first ten. It is the only digit that is not produced by any of the others—with the exception of one, of course—and that does not produce any other: six, eight, nine and ten being produced by the primes two, three and five, and all being produced by one, the unit.

"On which account," one ancient philosopher concluded, "other philosophers liken this number to Victory, who had no mother, and to the virgin goddess, whom the fable asserts to have sprung from the head of Jupiter: and the Pythagoreans compare it to the ruler of all things."

If he had been less of a numerologist and more of a mathematician, he might have pointed out other more significant ways in which seven is unique among the first ten numbers. Seven, for instance, is the only prime among the digits that is not one more than a power of two: two is $2^0 + 1$; three is $2^1 + 1$; five is $2^2 + 1$; but seven is one less than a power of two, $2^3 - 1$. The regular polygon with seven sides is the first with prime number of sides that cannot be constructed by the traditional methods of straightedge and compass alone.

One of the most interesting dramas in the theory of numbers is the discovery of a relationship between these last two apparently unrelated characteristics of the number seven. It is a story that is studded with some of the greatest names in mathematics.

For the beginning we must go back again to the Greeks. To the Greeks, as we have pointed out, numbers were also shapes. Each individual number was thought of as a polygon "with as many angles as units": three, a triangle; four, a square; five, a pentagon; six, a hexagon; seven, a heptagon; and so on. This interest in the shape of numbers extended even to their construction.

The Greeks were especially fond of limiting their constructions to those that were possible with straightedge (a ruler without any markings upon it) and compass alone and by proved principles of geometry. Their most famous construction problems were the trisection of an angle, the doubling of a cube and the squaring of a circle. All of these are now known to be impossible when the tools of construction are limited, as the Greeks limited them, to straightedge and compass. Without that limitation all are possible.

(Even though these famous constructions have been *proved* impossible, it is a rare student who comes to geometry for the first time without trying to make at least one of the constructions (the trisection of the angle is the most popular) and dreaming of achieving mathematical immortality when he should be studying Euclid's *Elements*.)

The problem of constructing regular polygons with straightedge and compass alone is somewhat different in that it is possible to construct some polygons, impossible to construct others. The man who determined the criterion for such constructibility was well on his way to mathematical immortality with his discovery before he was nineteen years old. But that was long after the Greeks.

The Greeks were easily able to construct a triangle and a square within a circle. Constructing a pentagon was somewhat more complicated but, thanks to the Pythagoreans, it was still possible under the classic restrictions. By bisecting the sides of these constructions they produced the other polygons with which we are most familiar—the hexagon, the octagon, the decagon and the dodecagon, all with equal sides and equal angles. But they could not, again under the classic restrictions, construct a regular seven sided polygon, a heptagon. They stopped, defeated, at seven. Were there more constructible polygons? If there were more, was their number finite or infinite? These questions remained unanswered for two thousand years. In that time no one constructed—with straightedge and compass alone—a regular prime polygon with sides numbering more than five.

The opening act of the drama of the constructible polygons had taken place in Greece. Act II, scene 1, was laid in France; scene 2, in Russia. At the time, no one in the audience suspected that the second act was even connected with the first. In it the leading role was played by Pierre Fermat, and the role he played was that of a very great mathematician being dead wrong.

Fermat was concerned, not with the constructible polygons, but with a particular form of number that he believed to be always prime. In the theory of numbers the search for a form that will invariably generate primes has been intensive. The only plausible conjecture anyone has made on the subject was made by Fermat. As it happened, his conjecture was false, and the numbers that still bear his name are a permanent reminder of his mistake.

It was the great mathematician's belief that numbers of the form $2^n + 1$ when n is a power of two were, without exception, prime. The first few numbers of the form $2^{2^t} + 1$ are certainly prime:

$$2^{2^0} + 1 = 3,$$
$$2^{2^1} + 1 = 5,$$
$$2^{2^2} + 1 = 17.$$

Fermat himself tested and found prime the next two of the form, 257 and 65,537. These are usually represented by a capital "F" and a subscript having the value of the respective power of two involved as F_3 and F_4. But testing F_5 was beyond even Fermat. In spite of its tidy representation by a capital letter and a single digit subscript, F_5 is a number that runs into the billions:

$$F_5 \text{ or } 2^{2^5} + 1 = 4,294,967,297.$$

Fermat made many attempts to find a factor for F_5 ("...j'ai exclu si grande quantité de diviseurs par demonstrations infaillibles," he wrote in 1640) and came to the conclusion (although being a mathematician he never went beyond "I think") that F_5, like the five numbers of that form that precede it, was prime and that all subsequent numbers of the form $2^{2^t} + 1$ were prime. These numbers are now permanently known as the Fermat numbers.

Some might consider the fact that the first five numbers of a particular form are prime a verification that all numbers of that form are prime, especially when the sixth is in the billions. For a mathematician, however, a sampling of any number of numbers is not enough to make a final statement about *all* numbers.

(In sciences other than mathematics a sampling must often serve as verification of a hypothesis. Mathematicians, who by the nature of their science can usually prove or disprove a hypothesis with finality, have a smug little joke that they call "The Physicist's Proof That All Odd Numbers Are Prime." The Physicist, so the joke goes, starts out by classifying one as prime because it is divisible only by itself and one. Then three is prime, five is prime, seven is prime, nine—divisible by three? well, that's just an exception—eleven is prime, thirteen is prime. Obviously all odd numbers with the exception of nine are prime.)

A mathematician's statement must be proved. To prove his, Fermat would have had to show that every number of the form $2^{2^t} + 1$ *must be prime*. To disprove it, someone had only to show that just one number of the form $2^{2^t} + 1$ is divisible by a number other than itself and one.

This is exactly what someone did, but not until almost exactly a century after Fermat made his statement about "si grande quantité de diviseurs" that he himself had tried on F_5. This someone was an equally great mathematician—Leonhard Euler, then mathematician at the court of St. Petersburg. Euler, as we have mentioned before, did not like to see mathematical questions lying around unanswered. Did the form $2^{2^t} + 1$ invariably generate primes, as Fermat had conjectured that it did, or did it not? Answering this question with finality in the negative could be as simple as finding a divisor of F_5, and this is what Euler set out to do.

He first determined that a factor of F_5, if such existed, would have to be a number of the form $2^{5+1}k + 1$, or $64k + 1$. With this discovery he greatly simplified the problem of testing the primality of F_5. Only certain possible primes of the form $64k + 1$ needed to be tried. The first few are 193, 257, 449, 577 and 641. As it happens, it was 641 that neatly divided F_5, or 4,294,967,297, and settled for all time that Fermat had been wrong. The form $2^{2^t} + 1$ does not invariably generate primes.

By all rights the curtain should have been rung down on the Fermat numbers. But it wasn't. These numbers, whether or not they are prime, are still very interesting. Like the powers of two from which they are formed, they are obviously infinite. Yet in all the infinitude of natural numbers there is not one number that divides *more than one* of the infinitude of Fermat numbers. This means that every one of the Fermat numbers (there being an infinite number of them) has a prime factor that not one of the others has.[1]

Mathematicians went right on looking for primes among the Fermat numbers after Euler had shown that all of them could not possibly be prime. There was still a mathematically interesting question to be answered. Were there any primes *at all* among the Fermat numbers beyond F_4? Or had it been the great mathematician's ill fortune that the only primes in the infinitude of numbers of the form $2^{2^t} + 1$ are the first five? In the second millennium the search continues, futile to date.

The third act of the drama took place in Germany, in 1801, with the publication of a small book. Two thousand years after the Greeks and a century and a half after Fermat, it brought to the stage again the five Fermat numbers, this time, to everyone's surprise, in the company of the ancient problem of the constructible polygons.

---

[1]This fact has been used for another neat proof of Euclid's theorem that the number of primes is infinite. The proof is by George Polya, whose very useful and untechnical little book, *How to Solve It*, is highly recommended to the reader.

The name of Carl Friedrich Gauss, the very young author of the book, is the only name in the theory of numbers that outshines the other names in this drama. Gauss was one of the three greatest mathematicians who ever lived (Archimedes and Newton being usually named with him), but in the branch of mathematics that is the theory of numbers no name is coupled with his. The small book published in 1801, when Gauss was twenty-four, was titled *Disquisitiones Arithmeticae*. Most of the work in it was done by him between the years of eighteen and twenty-one, the most profitable of his many profitable years. The *Disquisitiones Arithmeticae* is credited with systematizing the then completely unsystematized theory of numbers and marking out a path that other, lesser men were to follow gratefully.

It is appropriate, as we shall see, that Gauss took up the ancient problem of the constructible polygons in the seventh section of the *Disquisitiones*. This problem was not one that anybody expected to find in a book on the theory of numbers, for since the time of the Greeks it had been considered a problem in geometry. When it was solved, however, it was solved by an arithmetician who attacked it with algebra and found the answer in arithmetic.

Starting from the fact that the only constructible lengths are those that can be expressed algebraically using the four basic operations of arithmetic and square root, Gauss was able to show that a polygon with a prime number of sides can be constructed only if the prime is a prime of the form $2^{2^t} + 1$, and none other—in short, the favored primes of Pierre Fermat.

In general then, a regular polygon of n sides can be constructed with straightedge and compass alone only when n is a power of two, or a Fermat prime, or the product of a power of two and distinct Fermat primes.

With the general solution of the problem Gauss had added to the list of the basic constructible polygons just three:

> the regular polygon with 17 sides (F_2),
> the regular polygon with 257 sides (F_3),
> the regular polygon with 65,537 sidees (F_4).

Gauss, whose later mathematical achievements were numerous, was always very proud of this one, made when he was just eighteen years old. Supposedly it was the discovery that decided him between a career in philology and one in mathematics. He even suggested that a polygon with seventeen sides should be inscribed on his gravestone, as the sphere and circumscribed cylinder (suggesting the formula for the volume of the sphere) decorated Archimedes' tomb. But the stone mason objected that he could not carve a 17-sided polygon that would not look more like a circle. Whether Gauss ever made such a suggestion, he did point out, after his solution of the problem:

"There is certainly good reason to be astonished that while the division of the circle in three and five parts having been known already at the time of Euclid, one had added nothing to these discoveries in a period of two thousand years and that all geometers have considered it certain that, except for these divisions and those that may be derived from them..., one could not achieve any others by geometric constructions."

There is no seventeen-sided polygon on Gauss's gravestone, but one does appear on the monument erected to him in his native town of Brunswick.

But even Gauss did not answer the question whether the polygon of 65,537 sides (F_4) is the last that is constructible by the Greek requirements of straightedge and compass alone. This is a question that can be answered only when certain questions concerning the Fermat primes are answered. Is F_4 the last prime of the form $2^{2^t}+1$? If there are more, which seems increasingly doubtful, is their number finite—or infinite?

Since the time of Fermat a great many mathematical man-hours have been expended on these questions. The publication of the *Disquisitiones Arithmeticae*, giving as it did a new significance to the Fermat primes, made the answers even more interesting. Since Fermat made his conjecture, all that has been learned in the intervening centuries is that all subsequent Fermat numbers that have been tested are composite.

There are three methods for determining whether a Fermat number is composite. The first is a test similar to the Lucas test mentioned in "Six." As a result some Fermat numbers have been known to be composite years before a factor has been found. The numbers F_7 and F_8, for instance, were shown to be composite in 1905 and 1909 but factors were not found until 1970 and 1980. No factor has yet been found for F_{14} although it has been known to be composite since 1963. The test mentioned above, which was first stated in 1877, was used in 1987 as part of a long-term test of the hardware reliability of the Cray-2 supercomputer. In that year Jeff Young and Duncan A. Buell established the composite character of F_{20} after a total of approximately ten CPU days on the Cray-2. They concluded that to determine the character of F_{22}, which was at that time the smallest Fermat number of unknown character, they would need a little more than one hundred and sixty CPU days. Since that time, however, F_{22}, F_{23}, F_{24}, F_{28} and F_{31} have all been shown to be composite. But while we now know that F_{20}, F_{23} and F_{24} are composite and therefore the product of a unique combination of prime factors, we do not know (as we do not know in the case of F_{14}) even one of their factors.

(As in the case of the Mersenne primes, we must refer the reader to the internet for more up to date information.)

Incidentally, the test mentioned above can be used only for "comparatively small" Fermat numbers. Large numbers of that class must be attacked with a formula similar to the one used by Euler when he disposed of Fermat's conjecture by showing that F_5 is composite.

The two other methods of testing a Fermat number for primality are either *to find a single factor* of the number or *to factor the number completely*—the latter, of course, being almost always much more time consuming.

The Fermat numbers that have been tested do not include, however, all those that are below the largest Fermat number known to be composite. The explanation for this phenomenon is simply that the easiest numbers to establish as composite are those that have a prime factor among the first primes that would be tried as factors. For this reason, to bring the example down to a less olympian numerical level, it is much easier to factor a number like 14,997 than a smaller number like 8,633. The first prime that divides 14,997 is 3, but the first prime that divides 8,633 is 89.

We can find a somewhat similar example among the Fermat numbers in the case of F_{73}, the character of which was known as early as 1905. At the time of the invention of the computer, F_{73} was still the largest known composite Fermat number. In fact, it is probably the largest number the character of which was investigated in the pre-computer age. The number represented by F_{73} is so large that if it were printed in the decimal system in standard type and in standard-size volumes, all the libraries in the world could not hold it.[2] Fortunately, however, it and even larger Fermat numbers do not have to be written out in the decimal system before they can be factored. If a number F_t can be factored, the factor must be a number of the form $2^{t+2}k + 1$—an improvement over the $2^{t+1}k + 1$ mentioned earlier in connection with Euler, since it eliminates even more primes as possible factors. In the case of F_{73} this means a number of the form $2^{75}k + 1$. There are good mathematical reasons for not trying 1, 2, 3 or 4 for the values of k. So instead we take 5 and try $2^{75} \times 5 + 1$ as the first possible factor of F_{73}. It is indeed, as it turns out, the smallest prime factor. Thus, with very little work, the composite character of F_{73} was determined early in the twentieth century.

It is odd and interesting that among the Fermat numbers now known to be composite, eight have factors of the form $5 \times 2^n + 1$. These include the smallest composite Fermat number (F_5) and the largest ($F_{23,471}$) known to be composite. In two other cases there is a factor of the form $3 \times 2^n + 1$ and in four cases, a factor of the form $7 \times 2^n + 1$. (The value of n must be at least $t + 2$ for a factor of F_t.) Listed below are the cases involving these small primes. Not unexpectedly, the discovery of these numbers, unlike the discovery of other composite Fermat numbers, took place chronologically.

Oddly enough, while the complete factorization of F_{11} in 1988 passed virtually unnoticed in the nonmathematical world, the factorization two years later of F_9, billed by mathematicians interested in such things as one of "The Ten Most Wanted Numbers," attracted media attention from coast to coast. The *New York Times* headlined the story as GIANT LEAP IN MATH and noted that it was "an advance that could imperil secrets." This special attention was due

[2]We have taken our estimate of the size of F_{73} from W. W. Rouse Ball's *Mathematical Recreations and Essays*, a classic in its field.

F_t	Factor	Date of Discovery
F_5	$5 \times 2^7 + 1$	1732
F_{12}	$7 \times 2^{14} + 1$	1877
F_{23}	$5 \times 2^{25} + 1$	1878
F_{36}	$5 \times 2^{39} + 1$	1886
F_{38}	$3 \times 2^{41} + 1$	1903
F_{73}	$5 \times 2^{75} + 1$	1905
F_{117}	$7 \times 2^{120} + 1$	1956
F_{125}	$5 \times 2^{127} + 1$	1956
F_{207}	$3 \times 2^{209} + 1$	1956
F_{284}	$7 \times 2^{290} + 1$	1956
F_{316}	$7 \times 2^{320} + 1$	1956
$F_{1,945}$	$5 \times 2^{1,947} + 1$	1957
$F_{3,310}$	$5 \times 2^{3,313} + 1$	1979
$F_{23,471}$	$5 \times 2^{23,473} + 1$	1984

mostly to the fact that the factoring was a group project, utilizing several hundred mathematicians and about a thousand computers. The combination of human and computer power "brought the number in" speedily enough to show that the method is a real threat to "public codes." In these codes messages can be encoded using some very large number of a hundred digits or more that does not have to be secret because the messages can be decoded only by someone who knows the prime factors of that number.

"For the first time," announced the spokesperson for the group that did the complete factoring of F_9, a number of 155 digits, "we have gotten into the realm of what is being used in cryptography... it is impossible to guarantee security."

Fifty years ago, in the first edition of this book, we wrote: "It is not likely that the character of F_{13} [then the largest untested Fermat number] will be determined in the near future. Nor does it seem likely that the general question of whether the Fermat primes are finite will be determined soon."

We were wrong on the first count but right on the second.

The question of the constructible polygons is still open. Gauss was able to tell us that the regular polygon of seventeen sides (F_2) can be constructed by straightedge and compass alone; but even he, great as he was, was not able to tell us whether the polygon with 65,537 sides (F_4) is the last.

It seems unlikely that ten years from now there will be still another "Anniversary Edition" of *From Zero to Infinity.*

What is more than likely, in fact certain, is that there will be larger and still larger and still larger Fermat numbers that will have been proved composite. It has seemed, therefore, pointless to include in this edition information "at this time" regarding the largest Fermat numbers that have been factored.

What is less likely but not impossible is that some mathematician, perhaps still unborn, will come forth with A PROOF that there are no more Fermat primes beyond the first five that Fermat himself knew.

So the story of the constructible polygons and of the Fermat primes does not end, but merely stops. It is a story studded with some of the greatest names in mathematics, but there is still a place in it for another name.

· A CHALLENGE ·

Mersenne and Fermat numbers have much in common besides the fact that they both bear the names of men who guessed wrong. Mersenne numbers are of the general form

$2^n - 1$; Fermat numbers, of the form $2^n + 1$. Each form will produce primes only for certain limited values of n, and not always for these. For all other n it is not even necessary to write a number out to show that it can be factored. The problem stated below will make the challenge of both Mersenne and Fermat numbers more understandable.

Problem: For any positive integer s, $x^s - 1$ is algebraically divisible by $x - 1$. Similarly, if s is odd, $x^s - 1$ is divisible by $x + 1$.

$$x^2 - 1 = (x - 1)(x + 1)$$

$$x^3 - 1 = (x - 1)(x^2 + x + 1)$$

$$x^3 + 1 = (x - 1)(x^2 - x + 1)$$

It follows that $2^{rs} - 1 = (2^r)^s - 1$ is divisible $2^r - 1$ for all s and that $2^{rs} + 1 = (2^r)^s + 1$ is divisible $2^r + 1$ if s is odd. The following are special cases of the identities written above:

$$255 = 2^8 - 1 = (2^4 - 1)(2^4 + 1) = 15 \times 17,$$

$$511 = 2^9 - 1 = (2^3 - 1)(2^6 + 2^3 + 1) = 7 \times 73,$$

$$513 = 2^9 + 1 = (2^3 + 1)(2^6 - 2^3 + 1) = 9 \times 57.$$

Using the facts stated, the reader may enjoy finding some divisors of $2^{12} - 1$ (4,095) and of $2^{12} + 1$ (4,097). He may also consider in what cases the above rule does not give any factors of $2^n - 1$ or of $2^n + 1$ and try to draw a conclusion about when these numbers may be prime.

The number $2^{12} - 1 = 4,095$ is divisible by $2^2 - 1 = 3$, $2^3 - 1 = 7$, $2^4 - 1 = 15 = 3 \times 5$ and $2^6 - 1 = 63 = 3^2 \times 7$. A a matter of fact, $4,095 = 3^2 \times 5 \times 7 \times 13$.

The number $2^{12} + 1 = 4,097$ is divisible by $2^4 + 1 = 17$. Actually, $4,097 = 17 \times 241$.

In general, we can find a divisor, other than itself and one, of $2^n - 1$ unless n is prime and of $2^n + 1$ unless n is a power of 2. Hence, Mersenne numbers are numbers of the form $2^n - 1$ where n is prime; Fermat numbers are numbers of the form $2^n + 1$ where n is a power of 2. As we have seen in "Six" and "Seven," even with these limitations they are not always prime. That is why they present so great a challenge to the mathematician until he can prove whether they are finite or infinite.

· EIGHT ·

The most interesting thing about the number eight is that it is a cube ($2 \times 2 \times 2$), and the cubes are interesting and tough numbers. Since the time of the Greeks, who gave them their solid 3-D name, these numbers, which are the products of triple multiplication of the same number, have furnished the higher arithmetic with some of its most difficult problems. None has equaled in difficulty the problem that is today very simply the problem of the cubes. In that problem's history, the number eight, in addition to being itself a cube, has been a very significant number.

There are two questions that are usually asked about any group of numbers, and these have of course been asked about the cubes:

> How can the cubes be generally represented in the terms of the other natural numbers?
>
> How can the natural numbers be represented in the terms of cubes?

An answer to the first question dates from the early Christian era. It is usually credited to Nicomachus, whose

Introductio Arithmetica in the first century A.D. was the first exhaustive work in which arithmetic was treated independently from geometry. Cubical numbers, Nicomachus stated, are always equal to the sum of successive odd numbers and can be represented in this way:

$$1^3 = 1 = 1,$$
$$2^3 = 8 = 3 + 5,$$
$$3^3 = 27 = 7 + 9 + 11,$$
$$4^3 = 64 = 13 + 15 + 17 + 19,$$
$$\ldots .$$

An answer to the first question about the cubes was easy to find. (There may of course be other answers.) Answering the second question, of the general representation of the natural numbers in terms of cubes, was very difficult; and the answer when found inconsiderately posed a new, different and much more difficult question about the cubes.

When we speak of "representing" one group of numbers in terms of another, we may mean either by multiplication or by addition. It seems natural to think of the primes in terms of multiplication, and the integers are hence generally represented as the product of primes.[1] On the other hand, it seems natural to think of the cubes, like the squares, in terms of addition; the integers are then represented as the

[1]When mathematicians begin to think of numbers as the *sum* of primes they get into fantastic difficulties. In 1742 a Prussian mathematician named Christian Goldbach (1690–1764) offered what is now known as Goldbach's conjecture: every even number greater than four is the sum of two primes. No one doubted this postulate, but it was not until 1931 that a mathematician was able *to prove* that every even number is the sum of not more than three hundred thousand primes. Since then, it has been proved that every sufficiently large odd number is the sum of not more than three primes; hence every sufficiently large even number, of not more than four.

sums of squares, cubes, biquadrates and the other higher powers.

Obviously some integers require fewer cubes than others for representation as the sum of cubes. A number that is itself a cube, like eight, requires only one: 2^3. A number like twenty-three, which can be represented only in terms of the three smallest cubes since $3^3 = 27$, requires nine cubes for representation: $2^3 + 2^3 + 1^3 + 1^3 + 1^3 + 1^3 + 1^3 + 1^3 + 1^3$. Eight, however, like twenty-three, can also be said to be the sum of nine cubes since to 2^3 we can add 0^3 eight times for a total of nine cubes.

It is apparent then that *if there is a number that requires the most cubes for representation*, all numbers can be represented by that many cubes with the addition of the necessary 0^3s. But there was no assurance that there was such a number. The requirements for cubical representation might increase as the numbers themselves increase.

There was no serious attempt to answer the second question about the cubes until 1772. In that year a similar question about the squares, after unbelievable difficulty, had at last been answered with proof. There is no better example in number theory of the fact that it is easier to state a truth than to prove it. The Four Square Theorem states that every natural number can be represented as the sum of four squares. A little computation with the smaller numbers suggests that this is quite probably true. It is a theorem that is thought to have been familiar to Diophantus. Certainly it was stated by the translator through whom Fermat became familiar with the problems of Diophantus. It was then restated as part of a more general theorem and proved by Fermat. (This was the theorem that we met in "Five" to the effect that every number is either triangular or the sum of two or three triangular numbers; *square or the sum of two, three or four squares*; pentagonal or the sum of two, three, four or five pentagonal numbers; and so on). Although Fermat remarked that no proof ever gave him more pleasure, as usual the margins of his Diophantus were too small and the proof died with Fermat. Euler than tackled the problem of proving the part of the theorem pertaining to the squares, devoted in fact forty years, off and on, of his long life to it, without success. Eventually, though, in 1772, with the help of much of the work Euler had already done, the Four Square Theorem was proved by Joseph Louis Lagrange (1736–1813), the man Napoleon called "the lofty pyramid of the mathematical sciences." A few years later Euler brought forth a more simple and elegant proof than Lagrange's of the theorem that had caused him so much difficulty, and it is now the proof generally followed.

With a history like this behind its "twin" in the squares, it did not seem likely that the question of how many cubes are necessary and sufficient to represent any number as the sum of cubes would be easy to answer.

The year 1772, in addition to being the year of the long-sought proof of the Four Square Theorem, offered another incentive for trying to answer the question of cubical representation of numbers. Edward Waring had proposed, without proof, a theorem that went on—and on—from where the four square theorem left off. *Every number*, Waring suggested, *can be expressed as the sum of four squares, nine cubes, nineteen biquadrates, and so on through an infinitude of higher powers*. We met Waring in "Three" as the Cambridge professor who published John Wilson's unproved test for primality. Waring was something of a prodigy, being appointed to the faculty at Cambridge before he had obtained the necessary M.A. degree; as a result it had to be awarded to him by royal mandate. During his lifetime he was described as "one of the strongest compounds of vanity and modesty which the human character exhibits." ("The former, however," the writer added, "is his predominant feature.")

We will not at the moment go into the history of Waring's general theorem. It was Waring's good luck, not his fault, that it turned out to be "one of those problems that has started epochs in mathematics." (The words are E. T. Bell's.) As "Waring's problem" the theorem has paid off with an immortality in mathematics that Waring the mathematician never earned. (Oddly enough, in the summary of Waring's life in the *Dictionary of National Biography*, his mathematically famous problem is not mentioned.)

At this point we are concerned less with the general theorem than with Waring's choice of nine as the number of cubes necessary and sufficient to represent every number as the sum of cubes. That nine was quite probably the correct choice could have been suggested to him by a little paper and pencil work. If we start out to represent every number as the sum of cubes, we will find by the time we reach one hundred that none requires more than nine, and only 23, as we have already mentioned, requires as many as nine. If we continue past one hundred, we will find that there is not another number after 23 that requires as many as nine cubes until we get to 239.

It was probably on just such paper and pencil work that Waring based his statement that every number can be represented as the sum of nine cubes. It was a good guess, but nothing more. As we have already seen, there is no assurance implicit in paper work, no matter how far up into the numbers it is continued, that there are not numbers that require more than nine cubes for representation even though we never find them. Nor is there any assurance that the number of cubes required does not tend toward infinity as the numbers themselves do. This is what happens when we try to represent all numbers as the sum of powers of two, there being no fixed numbers of powers of two that will be sufficient for representation of all numbers.

The first step in proving the portion of Waring's theorem that deals with the cubes was to prove that there actually exists a finite number of cubes by which every number can be represented; in short, that the number of cubes required does not tend toward infinity. The mathematical symbol selected for this finite number of cubes, if such there was, was $g(3)$. By implication Waring had stated that such a $g(3)$ existed and was nine; that $g(4)$, the finite number of biquadrates necessary for representation of all numbers, was nineteen; and so on. Unless there existed a g for each power after $g(2)$, Waring's theorem had no meaning.There was no need to prove the existence of $g(2)$, since Lagrange had already proved (by proving the Four Square Theorem) that $g(2) = 4$.

It was not until 1895, more than a century after the publication of Waring's theorem, that even the existence, let alone the value, of $g(3)$ was established. At that time it was proved that every number can be represented as the sum of seventeen cubes. This meant that seventeen is sufficient; that the number of cubes required to represent any number, no matter what its size, can never be more than seventeen. Although it had not been proved that *the smallest possible number* of cubes necessary to represent all numbers was seventeen, it had been proved that seventeen was a bound to the number necessary: in short, an estimated value for a finite $g(3)$.

This was a great step because it disposed of the possibility that the cubes necessary to represent every number might, like the numbers themselves, continue to increase. Now $g(3)$ could be replaced by $G(3)$, the number of cubes actually required to represent all numbers as the sum of cubes.

For the next sixteen years mathematicians whittled away at seventeen, reducing the number of cubes sufficient and necessary for representation of every number as the sum of cubes from seventeen to sixteen to fifteen... and finally to nine. This conclusion was reached exactly one hundred and thirty-nine years after Waring had stated that every number can be represented as the sum of nine cubes.

Someone not familiar with the problems posed by the natural numbers might be inclined to think it a testimonial to Waring's brilliance to have recognized intuitively what it took his fellow mathematicians well over a century to come to by investigation and proof. This is not the case. For one of the characteristics of the natural numbers—perhaps their most interesting—is that some of the most easily guessed relationships among them are the most difficult to prove.

G. H. Hardy, who devoted a great deal of his time to Waring's problem, commented to this effect as follows:

"No very laborious computations would be necessary to lead Waring to a highly plausible speculation, which is all I take his contribution to the theory to be; and in the theory of numbers it is singularly easy to speculate, though often terribly difficult to prove; and it is only the proof that counts."[2]

It was at this point that the problem of the cubes, which had been difficult enough to take more than a century to

---

[2]Hardy is one of the most quotable of modern mathematicians, and it is only by a determined effort that we have refrained from quoting him even more often then we have. The reader is again recommended to his little book *A Mathematician's Apology*.

solve, became an incomparably more difficult—and more interesting—problem. In the year 1909 it was proved that the numbers requiring as many as nine cubes for representation are finite. Perhaps, as was generally suspected, 23 and 239 were *the only ones* in all the infinitude of numbers that require nine cubes.

What is the significance of the fact that only a finite number of numbers require as many as nine cubes? It is that there is some *last* number that requires nine cubes. From that number on, eight cubes are sufficient to represent all numbers.

We quote again from Hardy:

"Let us assume (as is no doubt true) that the only numbers which require 9 cubes for their expression are 23 and 239. This is a very curious fact which should be interesting to any genuine arithmetician; for it ought to be true of an arithmetician that, as has been said of Mr. Ramanujan, and in his case at any rate with absolute truth, that 'every positive integer is one of his personal friends.'[3] But it would be absurd to pretend that it is one of the profounder truths of higher arithmetic; it is nothing more than an entertaining arithmetical fluke. It is...8 and not...9 that is the profoundly interesting number."

[3]Srinivasa Ramanujan (1887–1920), the brilliant young Indian mathematician who died in 1920 at the age of thirty-two, has a colorful story that must certainly be included in a book on the interesting numbers. He was virtually self-taught until, as a clerk in the government service, having sent some of his mathematical work to several English mathematicians, he was brought to England by Hardy. For a few short years, the two men, Englishman and Indian, collaborated on some brilliant mathematical work. It is in the very fine memoir that introduces Ramanujan's collected papers that Hardy tells how, visiting his sick friend one day, he remarked that the number of the cab he had arrived in was 1,729, "not a very interesting number." Ramanujan replied promptly that on the contrary it was a very interesting number, being the smallest that can be represented as the sum of two cubes in two different ways ($1,729 = 10^3 + 9^3 = 12^3 + 1^3$).

With the new concept of the number of cubes that would be sufficient to represent all numbers from some point on (perhaps from 240 on), it was necessary to invent a new mathematical expression. Now $g(3)$, the number of cubes necessary to represent all numbers as the sum of cubes, was joined by $G(3)$, the number of cubes necessary to represent all numbers with a finite number of exceptions, perhaps only 23 and 239. It had already been established that $g(3)$ is nine; and since the numbers requiring as many as nine had been proved finite, it followed that $G(3)$ must be equal to or less than eight. In 1939 it was definitely established that 23 and 239 are the *only* numbers that require as many as nine cubes for their representation.

This distinction between "Little Gee" and "Big Gee," as they are sometimes called informally, was discovered through work on that phase of Waring's problem that dealt with third powers, but it had important implications for all other phases of the problem. The existence of $g(s)$ implies the existence of a $G(s)$, and the existence of $G(s)$ implies the existence of a $g(s)$. As a result, the mathematicians now found themselves with two problems for every one that they had had before: to determine a value for "Little Gee" for every power and another, the same or smaller, for every "Big Gee."

(The problem of "Big Gee" had never come up with the squares, for $g(2)$ and $G(2)$ are both four. Although all numbers except those of the form $4^m(8n+7)$ require only three squares for representation, numbers of this form are obviously infinite. There is, therefore, no number at which we can say, "From this point on all numbers can be represented as the sum of three squares." The question in regard to the fourth powers (the "biquadrates," as they were called in Waring's day) has also been answered: $g(4)$ is nineteen and $G(4)$ is sixteen.)

The problem of the cubes, as Waring had proposed it, had been solved; but as so often happens in the theory of numbers, the solution of one problem had produced another problem. As they had whittled away at $g(3) \leq 17$, mathematicians now began to whittle away at $G(3) \leq 8$. When tables of the actual cubical representation of the numbers up to 40,000 were examined, a curious fact emerged. There are among these only fifteen numbers that require as many as eight cubes for representation; seven are sufficient for all of the others (except of course 23 and 239, which we have already mentioned as requiring nine). The largest number requiring as many as eight cubes is 454. Between 454 and 40,000 *there are no other numbers requiring eight.*

Such paper and pencil work served to indicate, as it had before in the history of Waring's problem, a point of attack. Mathematicians set out to prove that the numbers requiring as many as eight cubes, like those requiring as many as nine, are finite. When they proved this—as they eventually did—the value of $G(3)$ was established as equal to or less than seven. This is where it stands at the date of writing. Yet there are indications in the same paper and pencil work that seven is quite probably not the final answer to the question.

In the table of numbers up to 40,000, there are only 121 for which as many as seven cubes are needed. The largest of these is 8,042. Between 8,042 and 40,000 *there are no numbers that require more than six cubes.* It is generally thought that from 8,042 on there are no numbers that require more than six cubes and that the value of $G(3)$ is probably equal to or less than six.

This is conjectured, not proved.

Yet when someone does prove, as eventually someone quite probably will, that $G(3)$ is equal to or less than 6, there is every indication in the table of numbers that this will not be the end of the matter either. Thousand by thousand, those numbers requiring as many as six cubes for representation become rarer. There are 202 numbers in the first thousand numbers requiring six. In the thousand numbers preceding a million, there is only one.

Someone may eventually prove that the numbers requiring as many as six cubes are also finite. Then the value of $G(3)$ will have narrowed to five or four, it already having been proved that there are an infinite number of numbers requiring four cubes for representation.

In the tables that have been made, it has been noted that there is a marked tendency for the numbers requiring five cubes to decrease as those requiring four increase. It is possible that eventually the five-cube numbers too will disappear; but if they do, it will be at a point far beyond the ability of man to make tables. This does not matter at all. The exact value of $G(3)$ can never be established by tables; it will have to be *proved.*

There is no question but that it will be very, very difficult to establish an exact value of $G(3)$. As we said in the opening of this chapter, eight and the other cubes are interesting and *tough* numbers.

· ANOTHER PROBLEM OF CUBES ·

Here is a problem of the cubes that, unlike the one we have been discussing, can be finally solved with a little paper and pencil work. Among all the numbers, there are just four that can be represented by the cubes of their digits—that are, in other words, equal to the sum of the cubes of their digits. What four numbers are they?

· ANSWER ·

$$153 = 1^3 + 5^3 + 3^3,$$

$$370 = 3^3 + 7^3 + 0^3,$$

$$371 = 3^3 + 7^3 + 1^3,$$

$$407 = 4^3 + 0^3 + 7^3.$$

· NINE ·

A great many things about the number nine and its relationships with the other numbers can be expressed by the equals sign; but there is one property of nine, known since antiquity and both interesting and useful, that cannot be so expressed. This is the fact that divided into any power of ten, nine always leaves a remainder of one. When, at the beginning of the nineteenth century, a notation very like that of the equals sign was at last invented to express this and other similar relationships, all of the numbers took on what might well be called a mathematical "new look." No single invention in the theory of numbers ever posed so many new and interesting questions. In the history of the number nine lies the seed for this sudden flowering.

In the days when computations were of necessity performed on counting boards, nine was commonly used as a check on the computation. Having completed his work and having before him on the board the beads of his answer, the person doing the computation needed to know if the answer was right. Thanks to nine, there was a very simple way for him to find out.

He had multiplied, let us say, 49,476 by 15,833 and had obtained the answer 783,353,508. His work no longer remained on the board, only his answer:

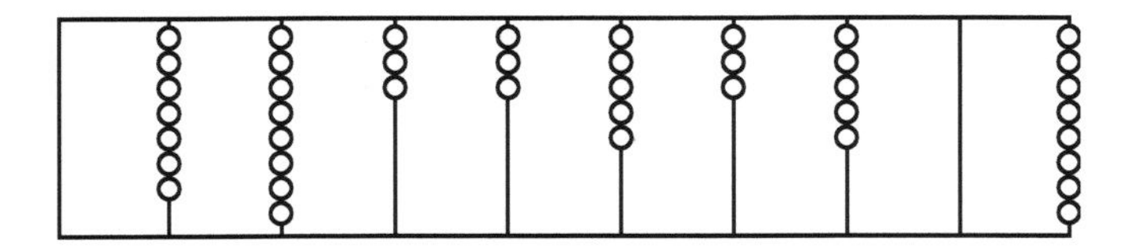

Since he knew that nine leaves a remainder of one when divided into any power of ten and since each bead on the board represented a power of ten, he proceeded to count the beads of the answer as he had previously counted the beads representing the numbers to be multiplied. Today we would add the digits:

$$1 + 5 + 8 + 3 + 3 = 20,$$
$$4 + 9 + 4 + 7 + 6 = 30,$$
$$7 + 8 + 3 + 3 + 5 + 3 + 5 + 0 + 8 = 42.$$

He then divided each of these sums by nine, noting only the remainders:

$$20 \div 9, \text{ a remainder of } 2,$$
$$30 \div 9, \text{ a remainder of } 3,$$
$$42 \div 9, \text{ a remainder of } 6.$$

If the computation has been correct, the remainders of the numbers being multiplied (reduced when necessary by further casting out of nines) will produce the remainder of the product. Since $2 \times 3 = 6$, he could go on with a fair amount of confidence to his next problem. (There was, however, always the possibility that digits might have been transposed in the answer, a common error that the check will not catch.)

This same check can be used for addition and subtraction as well as for multiplication. The sum of the same two numbers we multiplied will leave a remainder of five; their difference, a remainder of one. For checking division, we follow the standard rule that the dividend a should equal the divisor b multiplied by the quotient q plus the remainder r, or $a = b \times q + r$. But instead of using the whole number for this check, we cast out nines and use only the remainders.

$$\begin{array}{r} 3 \\ 15833\,\overline{)49476} \\ 47499 \\ \hline 1977 \end{array}$$

$$49{,}476 = 15{,}833 \times 3 + 1{,}977$$

or (when the nines are cast out)

$$2 \times 3 + 6 = 12 \text{ and } 1 + 2 = 3$$

This was the ancient computational check known as "casting out nines." It depends upon the fact we have already noted that nine when divided into one, ten, one hundred, one thousand or any other power of ten leaves a remainder of one. For this reason nine divides a number that is represented in the decimal system only if it also divides the sum of the digits of the number. If it leaves a remainder, this is the same remainder that would be left if the number itself were divided by nine.[1] By this ancient "Rule of Nine" we can say, with only the hesitation of the time that it takes us to add the digits and divide them, that such a number as 9,876,543,210, for example, is divisible by nine:

[1]A similar check exists with eleven, which alternately leaves a remainder of +1 or −1 when divided into the powers of ten (+1 for 1, −1 for 10, +1 for 100, −1 for 1,000, and so on). To check a computation by elevens, we alternately add and subtract the digits and then divide the total by eleven.

$$9+8+7+6+5+4+3+2+1+0=45,$$
$$4+5=9,$$
$$9\div 9=1.$$

The notation that was at last invented to express a relationship between any two numbers such as the one that exists between nine and the powers of ten is a beautifully simple one. It was invented by Gauss, "whose name," as E. T. Bell wrote, "lives everywhere in mathematics." The language of Gauss's *Disquisitiones Arithmeticae* is Latin, but the mathematical language is that of *congruence,* there used for the first time.

A *congruence* is a relationship similar enough to the relationship expressed by the equals sign to be very useful, different enough to be very interesting:

$$= \text{ equal to,}$$
$$\equiv \text{ congruent to.}$$

Gauss gave in the *Disquisitiones* the following definition:

> Two integers a and b shall be said to be congruent for the modulus m [from the Latin, "a certain small measure"] when their difference $a-b$ is divisible by m.

$$a \equiv b(\bmod m) \qquad 5 \equiv 1(\bmod 2)$$
$$84 \equiv 0(\bmod 6)$$
$$173 \equiv 8(\bmod 11)$$

Another way of saying the same thing is to say that a and b leave the same remainder when divided by m.

Although the notion of congruence may seem completely unfamiliar when we first meet it, it is not—in fact, it is very familiar. We base every day of our lives on a congruence relationship. When we say that today, for example, is Tuesday, we are saying that a certain number of days when divided by seven (the week) leaves a remainder of Tuesday.

The day of the week can be very accurately stated as a congruence if we utilize the astronomers' concept of the Julian day. To avoid the confusion that results from months and years of unequal lengths, the astronomers number the days consecutively from January 1, 4713 B.C., the beginning of the Julian era. By this numbering January 1, 1930, which fell on Wednesday, was Julian Day 2,425,978. With this information and the congruence relationship based on the modulus 7 we can compute on which day of the week January 1, 2000—25,567 days later—would fall:

$$\begin{aligned}\text{January } 1, 1930 &= \text{J.D. } 2,425,978 \equiv 2(\bmod 7)\\ &= \text{Wednesday},\\ \text{January } 1, 2000 &= \text{J.D. } 2,451,545 \equiv 5(\bmod 7)\\ &= \text{Saturday}.\end{aligned}$$

The general congruence upon which the ancient method of checking a computation by casting out nines rests is

$$10^n \equiv 1(\bmod 9).$$

This notation tells us at a glance that the difference between one and any power of ten is always divisible by nine. When, instead of looking merely at the powers of ten in relation to the number of nine, we look at *all* numbers for the same modulus we find that they fall into nine different groups:

$$\begin{array}{lcl}
0, \quad 9, 18, 27, 36, \ldots & \equiv & 0 \pmod 9 \\
1, 10, 19, 28, 37, \ldots & \equiv & 1 \pmod 9 \\
2, 11, 20, 29, 38, \ldots & \equiv & 2 \pmod 9 \\
3, 12, 21, 30, 39, \ldots & \equiv & 3 \pmod 9 \\
4, 13, 22, 31, 40, \ldots & \equiv & 4 \pmod 9 \\
5, 14, 23, 32, 41, \ldots & \equiv & 5 \pmod 9 \\
6, 15, 24, 33, 42, \ldots & \equiv & 6 \pmod 9 \\
7, 16, 25, 34, 43, \ldots & \equiv & 7 \pmod 9 \\
8, 17, 26, 35, 44, \ldots & \equiv & 8 \pmod 9
\end{array}$$

Every number falls into one of these nine groups, and no number falls into more than one group. With the congruence notation, as in the method of casting out nines, we are now able to treat all numbers as if they were just nine different numbers. A specially constructed multiplication table gives us all possible products for the modulus nine:

×	0	1	2	3	4	5	6	7	8
0	0	0	0	0	0	0	0	0	0
1	0	1	2	3	4	5	6	7	8
2	0	2	4	6	8	1	3	5	7
3	0	3	6	0	3	6	0	3	6
4	0	4	8	3	7	2	6	1	5
5	0	5	1	6	2	7	3	8	4
6	0	6	3	0	6	3	0	6	3
7	0	7	5	3	1	8	6	4	2
8	0	8	7	6	5	4	3	2	1

Using this table, the reader will find that such apparently dissimilar multiplications as 13×14, 4×32 and 22×41 give the same answer (mod 9); or, as we said earlier, when we cast out the nines, they will all leave a remainder of 2.

Each pair contains one number that is congruent to 4 (mod 9) and one number congruent to 5 (mod 9). The reader will note that 4×5 on the above multiplication table yields 2.

By performing the multiplications indicated, he will find that all three products are congruent to 2 (mod 9).

Just as we have looked at all numbers in relation to nine, we can look at them in relation to any number m that we choose and find that they will accordingly fall into one of m mutually exclusive groups. The most familiar way of doing this is according to the number two.

> An *even* number n leaves a remainder of 0 when divided by 2.
>
> An *odd* number n leaves a remainder of 1 when divided by 2.
>
> An *even* number n is one that is congruent to 0 (mod 2).
>
> An *odd* number n is one that is congruent to 1 (mod 2).
>
> Or an *even* number $n \equiv 0 \pmod{2}$.
>
> An *odd* number $n \equiv 1 \pmod{2}$.

The notation invented by Gauss in the *Disquisitiones* was at once so exact and so easily grasped that many theorems that were already known in other forms were promptly restated as congruences. A case in point is Wilson's theorem, which we have met earlier in "Three." The expression of this theorem as a congruence is today so usual that a mathematician, hearing that the author intended to introduce Wilson's theorem in "Three" but not to mention the congruence notation until "Nine," demanded, "But how can you even *state* Wilson's theorem until you have explained congruence?" Yet seven years before the inventor of the congruence notation was born, Wilson's theorem was stated as follows:

If p is a prime, then the quantity

$$\frac{1 \times 2 \times 3 \times \cdots \times (p-1)+1}{p}$$

is a whole number.

When young Wilson's teacher, Edward Waring, published this theorem in 1770, he commented: "Theorems of this kind will be very hard to prove, because of the absence of a notation to express prime numbers." It was in connection with this remark that Gauss commented sharply to the effect that mathematical proofs depend on *notions*, not on *notations*. Although today Wilson's theorem is almost invariably expressed in the notation of congruence as

$$(p-1)!+1 \equiv 0(\bmod p),$$

and although the simplest and most direct proof of Wilson's theorem (Gauss's own) is based on congruence, the notion still remains more important than the notation.

There is, nevertheless, in the history of congruence a strong argument for notations sharing importance with notions. The type of relationship that is expressed by the three parallel lines of the congruence has been known since the early centuries of the Christian era. There is even an equally brief way of noting it mathematically with the symbol "$|$" for "divides":

To say $m|(a-b)$ is the same as saying $a \equiv b$ (mod m).

Yet this long known type of relationship played no important part in the study of numbers until Gauss found a way of expressing it in a mathematically suggestive form. The three parallel lines of the congruence sign suggest the equals sign and remind us that congruences and equalities, both being equivalence relationships, have certain properties in common. We are familiar with these in equalities:

If a is any number, then $a = a$.
If $a = b$, then $b = a$.
If $a = b$ and $b = c$, then $a = c$.

These properties of the equality relationship are also properties of the congruence relationship:

If a is any number, then $a \equiv a \pmod{m}$.
If $a \equiv b \pmod{m}$, then $b \equiv a \pmod{m}$.
If $a \equiv b \pmod{m}$ and $b \equiv c \pmod{m}$, then $a \equiv c \pmod{m}$.

The similarities between equalities and congruences that are emphasized by the similar notation suggest that we attempt with congruences certain operations that work with equalities. We have already seen, in the process of casting out nines, how we can add, subtract and multiply numerical congruences as we do equations. We can also handle algebraic congruences very much as we handle algebraic equations. The results are usually interesting.

Consider, for instance, a fundamental problem of squares and primes:

> To find a square one less than a multiple of p where p is a given odd prime.

In the congruence notation this problem is stated more briefly:

Is $x^2 \equiv -1 \pmod{p}$ solvable?

Before we give the general solution of this problem, the reader might like to try to solve it for a few values of p: to find squares that are, respectively, one less than a multiple of the first few odd primes, three, five, seven, eleven and thirteen. He will find them for only two of these primes, but those he can find very quickly.

Now for the general solution of the problem.

It can be proved, although not here, that the only odd primes for which the congruence above is solvable are those that like five and thirteen are of the form $4n + 1$. In the congruence notation we say

$$x^2 \equiv -1 \pmod{p} \text{ is solvable only when } p \equiv 1 \pmod{4}.$$

Closely connected with this problem is a theorem that has the distinction of being the most often proved in the theory of numbers. That it can be approached in so many different ways speaks eloquently for its fundamental importance in number relationships. We mention it here because it is the finest example of that particular type of relationship among the numbers that is brought to the fore by the congruence notation.

The theorem, which is known as the Law of Quadratic Reciprocity, was called by Gauss himself *the gem of arithmetic*. Since at another time Gauss called mathematics *the queen of the sciences* and arithmetic *the queen of mathematics,* this puts the law of quadratic reciprocity at the very pinnacle of science.

The Law of Quadratic Reciprocity was known to mathematicians before Gauss. It was Euler who discovered it, but neither he nor anyone else proved it. Then at the age of eighteen, unaware of the work of Euler and others, Gauss rediscovered the law on his own. He found it immediately beautiful, but he was not immediately able to prove it. "It tortured me for the whole year and eluded the most strenuous efforts," he wrote. He proved it, at last, in a fittingly beautiful and simple form. At the time he was nineteen years old.

Having proved this gem of arithmetic, Gauss was still so fascinated by it that during the course of his lifetime he composed six more very different proofs. At the time of this writing the number of proofs of the Law of Quadratic Reciprocity is well over one hundred.

The "reciprocity" of the law is that which exists between two different odd primes p and q. The law states that for p and q the two congruences

$$x^2 \equiv q \pmod{p} \text{ and } x^2 \equiv p \pmod{q}$$

are both solvable or both not solvable unless both p and q are primes of the form $4n - 1$, in which case one of the congruences is solvable and the other is not.

We can see and admire the Law of Quadratic Reciprocity in action if we try to determine whether a particular congruence of the type to which the law applies is solvable; for example:

$$\text{Is } x^2 \equiv 43 \pmod{97} \text{ solvable?}$$

This is the same as asking ourselves whether there exists a square that is 43 more than a multiple of 97. Since not both primes, 43 and 97, are of the form $4n - 1$, we know by the Law of Quadratic Reciprocity that the congruence

$$x^2 \equiv 43 \pmod{97}$$

is solvable only if the congruence

$$x^2 \equiv 97 \pmod{43}$$

is also solvable. They stand or fall together: either *both* are solvable, or *both* are not solvable.

To determine the solvability of our second congruence, which will settle the solvability of the first, we proceed, since 97 is greater than 43, to reduce it by dividing 97 by 43 and obtaining as our answer

$$x^2 \equiv 11 \pmod{43}.$$

We now have a congruence in which both primes are of the form $4n - 1$. We know by the Law of Quadratic Reciprocity that

$$x^2 \equiv 11 \pmod{43} \text{ is solvable}$$

only if

$$x^2 \equiv 43 \pmod{11} \text{ is not solvable.}$$

We now proceed as before, since 43 is greater than 11, to reduce this second congruence. It reduces to a familiar one:

$$x^2 \equiv -1 \pmod{11}.$$

We recognize this as the same problem we met a few pages back: to find a square one less than a multiple of p when p is a given odd prime. We recall from the solution of that problem that the congruence expressed above is solvable only when the prime is of the form $4n + 1$. Since 11 is of the form $4n - 1$, the congruence is not solvable.

We can now work our way back to our original congruence.

Since $x^2 \equiv -1 \pmod{11}$ is not solvable, then by the Law of Quadratic Reciprocity$x^2 \equiv 11 \pmod{43}$ is solvable. Since $x^2 \equiv 11 \pmod{43}$ is solvable, $x^2 \equiv 97 \pmod{43}$ is solvable and therefore by the Law of Quadratic Reciprocity our original congruence $x^2 \equiv 43 \pmod{97}$ is solvable.

We have not actually found *a solution* to the congruence. (This is often more difficult than proving that a solution exists, but never so interesting.) As it happens, however, for this particular congruence we can find the numerical solution simply by inspection. The congruence

$$x \equiv \pm\, 25 \pmod{97}$$

means that any x that differs from a multiple of 97 by 25 will, when squared, be exactly 43 more than a multiple of 97. The lowest positive value for x is 25, and the reader may be interested in testing the original congruence for $x = 25$.

The solution to such a congruence looks quite a bit like the solution to an equation, but there is a significant difference. For an equation, such as

$$x^2 - 625 = 0,$$

which also has as its solution ± 25, there are only two values among the infinitude of the integers that when substituted for x will "work." These are $+25$ and -25.

On the other hand, for the congruence that we have just solved, although we say also that there are only two solutions, each of these solutions actually stands for an infinitude of numerical values that will satisfy the congruence

$$x^2 \equiv 43 \pmod{97}.$$

In this congruence x can be any number, *positive or negative*, that differs by 25 from a multiple of 97.

That we are able to speak of this infinitude of values simply as two numbers is indicative of the new perspective on the numbers that we gain from the notion of congruence. Normally when we look at the numbers we try to get as close as possible to them so that we can see the ways in which they are different from one another. When, however, we look at the numbers in terms of congruences, we move away from them. Suddenly they do not look as much different as they look alike. We are able, as in the solution to our congruence a few pages back, to see an infinitude of numbers as being the same. Because they are all congruent to one of a pair of numbers in respect to the same modulus, we can think of them, *not as an infinitude of different numbers, but as two.*

It is a thought-provoking transformation.

For if the numbers, so seemingly regular and predictable as they stretch out by ones to infinity, are capable of such a transformation, what else may they not be capable of?

· ONE FOR THE READER ·

Is the congruence $x^2 \equiv 2 \pmod{p}$ solvable?

With a knowledge of the solution of this congruence and the solution of the congruence $x^2 \equiv 1 \pmod{p}$, which we gave in this chapter, combined with the Law of Quadratic Reciprocity, it is possible to determine the solvability of any congruence $x^2 \equiv a \pmod{p}$.

Although it is not at all easy to prove under what conditions the congruence

$$x^2 \equiv 32(\text{mod} p)$$

is solvable, the reader may be able to guess them by actually testing the congruence for the first few squares

$$0, 1, 4, 9, 16, 25, 36, 49, 64, 81$$

and the first few odd primes

$$3, 5, 7, 11, 13, 17, 19, 23, 29, 31.$$

· NINE ·

· ANSWER ·

The congruence $x^2 \equiv 2 \pmod{p}$ is solvable when $p \equiv \pm 1 \pmod{8}$. It is solvable for 7, 17, 23 and 31 among the above primes. It is not solvable when $p \equiv \pm 3 \pmod{8}$, so it is not solvable for 3, 5, 11, 13, 19 or 29.

· EULER'S NUMBER ·

Everything about the natural numbers is *there*. All the relationships and interrelationships among them are inherent in the ordered unit-by-unit sequence that begins with zero and continues to infinity. The simple patterns of the surface are easy for anyone to guess but often difficult or even impossible to establish by proof. Subtler, more complicated patterns lie so deep that only the rarest minds glimpse them. Yet all are implicit when we begin 0, 1, 2, 3,. . . .

It is the more amazing then that one of the most stubbornly guarded secrets of the natural numbers—the general distribution of the primes—should be wrested from them by means of a number that is not at all *natural* in the same sense that they are.

Such a number is the one that mathematicians call "Euler's number" or, more simply, e. It is a number that cannot be expressed by any finite combination of the integers. A number that did not come into formal existence until nearly two thousand years after the Greeks began their numerical investigations. A number that, although it seems most unnatural, has a more intimate connection with nature than any one of the natural numbers.

The story of this very interesting number e and the way in which, with the knowledge of e, mathematicians were able to uncover the very deep, very important relationship of the infinitude of prime numbers to the infinitude of natural numbers is perhaps the most amazing in twenty-five centuries of number theory, and a fitting one to include in a book subtitled "What Makes Numbers Interesting."

The nearest we can come to an exact numerical representation of e is the famous factorial series:

$$e = 1 + \frac{1}{1!} + \frac{1}{2!} + \frac{1}{3!} + \frac{1}{4!} + \frac{1}{5!} + \frac{1}{6!} + \frac{1}{7!} + \frac{1}{8!} + \frac{1}{9!} + \cdots.$$

Although at first glance we may find it strange to express a number as the limiting sum of an infinite series, we actually do the same thing every day. The decimal .333333... is a familiar representation of just such an infinite series for the number $1/3$:

$$\frac{1}{3} = \frac{3}{10^1} + \frac{3}{10^2} + \frac{3}{10^3} + \frac{3}{10^4} + \frac{3}{10^5} + \frac{3}{10^6} + \frac{3}{10^7} + \cdots.$$

If we perform the indicated additions $(3/10 + 3/100 + 3/1,000 + \ldots)$ on a mental number line, we intuitively recognize that while we can get as close to the point $1/3$ as we wish, we are never going to hit it exactly nor can we exceed it. In the same way, as we sum more and more terms of the series for e, we get closer and closer to the exact value e, which is the limiting sum of that series just as $1/3$ is of the other.

From the factorial series for e that begins above we can also obtain a decimal representation of that number to as many places as we desire. We proceed in the following manner:

We take 1, then take 1 again and divide by 1, take the answer—which is of course 1—and divide by 2, take the answer (.500000) and divide by 3, take the answer and divide by 4, and so on. When we have divided by all the numbers up to and including 9, and have added all our quotients, we have the number e rounded off to six decimal places:

$$\begin{array}{r} 1.000000 \\ 1.000000 \\ 0.500000 \\ 0.166667 \\ 0.041667 \\ 0.008333 \\ 0.001389 \\ 0.000198 \\ 0.000025 \\ 0.000003 \\ \hline e = 2.718282. \end{array}$$

This process can continue without end, like the factorial series on which it is based. There is, however, a difference between the decimal representation of e and its representation as the limit of the series above. While we can always predict the nth term of the series above, which will be $1/(n-1)!$, we have no possible way of knowing in advance what digit will be in the nth place of the decimal representation of e. In this the number e differs from $1/3$ and the other rational numbers; for they can always be represented by decimals that at least after a certain point repeat in a regular, predictable pattern. The number e is thus nonrational, or *irrational*.

To gain even the most superficial knowledge of such a seemingly un-number-like number, we must examine e in relation to the many extensions of the idea of number that have been made since the days when the Pythagoreans built a philosophy and a religion—as well as a science—on the basis of the numbers 1, 2, 3, ...

The Pythagoreans believed that the universe was "ruled" by these numbers. Although they recognized that there were lengths, for example, that could not be measured in whole numbers, they were sure that they could always put a number to such lengths by using the ratios of whole numbers, like 1/3, 5/7, and so on. In other words, if all possible lengths were thought of as represented by points on some giant measuring line, there was among the whole numbers and their ratios a number for every possible point on that line. A death blow was dealt to this theory when, approximately four centuries before the birth of Christ, they discovered and proved that among the whole numbers and their ratios there is no such number or ratio that exactly measures the diagonal of the unit square:

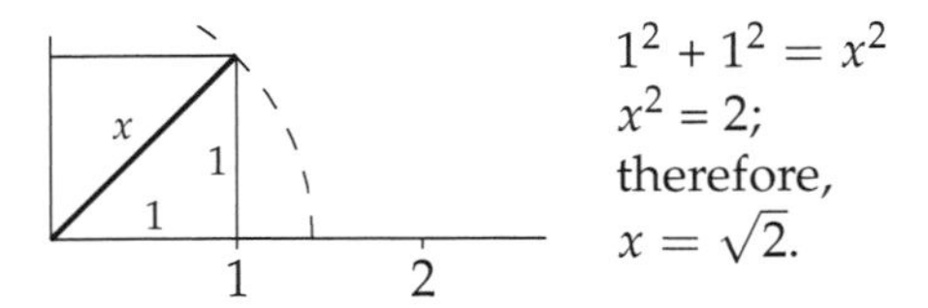

They proved this is true by first assuming that there was some ratio between two whole numbers a and b such that $(a/b)^2$ was equal to two, and then showing that such an assumption led logically to an impossibility and, therefore, must be false.

Since the Greeks had never really considered even the ratios to be *numbers*, it never occurred to them that a nonratio like the square root of 2 could possibly be a number. The conclusion they drew from their discovery was that there were lengths for which there were no numbers and that they would be better off studying geometry, where there would be no necessity to put *a number* to the diagonal of the unit square before they could use it in their mathematics.

(One of the most unusual things about the number *e* is that although we know it measures a definite distance from zero on the number line and although we can approach the point it represents to any degree of accuracy we care to demand, we cannot with the traditional instruments of mathematics actually produce a line segment of exact length *e*. This paradox would have been appreciated by the Greeks, who in the diagonal of the unit square were able to produce a line segment for which they could produce no number.)

It was the non-Greek development of algebra that at last brought mathematicians back to numbers. With the invention of zero and the negative numbers, they could think of the number line as extending indefinitely in both directions. The invention of decimals at the beginning of the sixteenth century gave them a numerical representation for every conceivable point on the line, including the length of the diagonal of the unit square. The negative numbers marked the lengths to the left of the origin (0); the positive numbers, the lengths to the right. The rational numbers, positive and negative, marked all those lengths that could be represented as ratios between integers, including the integers themselves. As decimals they either repeat (like 1/3 or 1/7) or terminate (like 1/5). The irrational numbers marked all other lengths, or points, and (like the square root of 3 and the square root of 7) are nonrepeating, nonterminating decimals.

It all seemed crystal clear. There was a number for every point on the line, and (by the definition of irrational numbers) there could not be a point for which there was no number, or a number for which there was no point. Because by this time mathematicians were beginning to use as numbers other quantities whose reality as numbers seemed more doubtful to them than even the square root of 2 had seemed to the Greeks, they christened the numbers that they could place in one-to-one correspondence with the points on a straight line *the real numbers*.

The numerically more doubtful quantities that we just mentioned depended upon the idea of "a fiction" that they called "i, the square root of –1." By the use of this fiction they had found that they could solve equations that were otherwise unsolvable even with all the real numbers they already had at their disposal. Such an equation is

$$x^2 + 1 = 0,$$

where it is clear that x^2 must be -1 and that x must be some number that when squared, or multiplied by itself, produces -1. They all agreed that there could be no such number since any positive or negative number *when multiplied by itself* yields a positive product. But they also agreed that it would be very useful mathematically if there were such an impossible number, so they began to use i to provide them with square roots for negative numbers. They called these *the imaginary numbers*.

The extension to the number i was the *last* one needed to provide a root for every algebraic equation.

Up until this time, in spite of all the new quantities being used as numbers, the quantity represented by the nonrepeating, nonterminating decimal 2.7182818... that we call e had not been picked out for any special attention. Being, however, one of the real numbers, the representative of a particular point, it was most certainly in existence there on the real number line, somewhere between 2 and 3, between 2.7 and 2.8, between 2.71 and 2.72, somewhere—but *there*.

It was not until the invention of logarithms at the beginning of the seventeenth century that the value represented by 2.7182818... was recognized as one of the most interesting of numbers, the base of the so-called *natural* logarithm.

The principle of logarithms, invented by John Napier, Laird of Merchiston (1550–1617), immeasurably reduced the burden of calculation with very large numbers by replacing multiplication by addition. (Napier himself was particularly interested in the problems of astronomical calculation.) This was one of those "Why didn't I ever think of that?" inventions, and the feeling was never better expressed than by Henry Briggs (1556–1631), professor of geometry at Oxford, who upon beholding the inventor of logarithms for the first time, marveled:

"My lord, I have undertaken this long journey purposely to see your person, and to know by what engine of wit or ingenuity you came first to think of this most excellent help in astronomy ... but, my lord, being by you found out, I wonder nobody found it out before, when now known it is so easy."

We are all familiar with how simple it is to calculate with exponents. We can, for instance, multiply perfect powers of a given number by adding the exponents, so that $10^{15} \times 10^{23}$ becomes merely a problem in addition, 15 + 23, to give us the product 10^{38}. Using the principle of logarithms, we can change *every* number into a power of the same base, so that while 10 is 10^1, a number that is not a perfect power of 10, like 11, is $10^{1.0414}$, approximately, and 12 is $10^{1.0792}$. The principle can also be extended to the other side of the decimal point so that 11.5, for example, is $10^{1.0607}$, again approximately.

Mathematically, we express what we have just done by two simple formulas that embody the principle of logarithms:

$$x = 10^y \text{ and } y = \log_{10} x.$$

The terminology of logarithms follows naturally:

$10 = 10^1$, and 1 is the logarithm of 10,

$11 = 10^{1.0414}$, and 1.0414 is the logarithm of 11.

Logarithmic calculations originally used the base ten, which seems to us most "natural." However, as we have seen earlier in this book, there is nothing mathematically natural about ten as a number base. Whatever naturalness it has follows, not from any property of the natural numbers themselves, but only from the fact that we are born with ten fingers. To mathematicians, doing analytic work, and to engineers, doing any kind of computation that involves calculus, it is much more natural to use e rather than ten as a base for logarithms. For this reason e is called the base of the *natural* logarithm.

For more than three hundred years, the ease of calculation with logarithms made them indispensable to people who worked in a practical way with numbers. With the invention and proliferation of computers, they and their pocket aid, the slide rule, have become artifacts of another time. (Even the counting board has outlived them!) But the natural logarithm is still as important as ever.

Most of the reasons for e being mathematically natural are too technical for a book of this type, but one example will serve to illustrate the mathematical naturalness of base e over base ten.

Let us say that we wish to choose a number for a base such that the logarithm of any small number $(1 + x)$ that differs from one by the very small amount x will be approximately equal to x itself. Such a logarithm greatly simplifies calculations with very small numbers. What number then can we choose for our base? The answer is not what we would expect. Surprisingly, it can be established with finality that the best possible number for a base meeting the requirement above is the (to us) unlikely number e.

The chart here shows the difference between logarithms to base e and base ten, respectively, for several numbers that differ from 1 by only a few hundredths:

	1.00	1.01	1.02	1.03	1.04	1.05
$\log_e$	0.0000	0.0100	0.0198	0.0296	0.0392	0.0488
but						
$\log_{10}$	0.0000	0.0043	0.0086	0.0128	0.0170	0.212

From this chart we can see that $\log_e 1.01$ is exactly equal (to four places) to the difference between 1 and 1.01 while $\log_e 1.02$, which is .0198, is almost equal to .02. On the other hand, $\log_{10} 1.01$ is .0043. Mathematically, we say that $\log_e(1 + x)$ is approximately equal to 1 times x, *or* x itself, for very small x while $\log_{10}(1 + x)$ is approximately equal to .43 times x. Since 1 is obviously a much more "natural" number to work with than .43, we use the natural logarithm to base e for calculation with very small numbers and dismiss 10 as the base of the common logarithm.

But the number e is not merely mathematically natural. Although its definition as a logarithmic base is removed from everyday base-ten life and its character as a number is different from that of any one of the natural numbers, it is a number most intimately a part of nature itself. The basic processes of life—growth and decay—are most accurately represented in mathematical terms by curves that spring from what mathematicians call *the exponential function,* or generally the curve determined by the equation $y = e^x$. Thus the number e is uniquely important in many different applications of mathematics: probability and statistics, biological and physical sciences, ballistics, engineering, finance.[1]

Long before it was given the name by which we know it best today, the number 2.7182818... was recognized as the base of the mathematically natural logarithm. The young Euler—he was just twenty-one years old and at the court of St. Petersburg—first suggested the alphabetical name in a paper entitled "Meditation upon Experiments made recently on the firing of Cannon."

"For the number whose logarithm is unity," he stated, "let e be written... ."

Although e is often referred to as "Euler's number" and will probably always suggest the first letter of the great mathematician's name, it was probably selected by him for quite another reason. It is the vowel immediately following a,which he customarily used to signify the general logarithmic base.

"Euler's number," however, is a most appropriate name for the number e. Probably never in the history of mathematics has a great affinity existed between one man and one number. Euler calculated e to twenty-three places, surely a labor of love at that time:

$$e = 2.71828182845904523536028\ldots.$$

He found several very simple representations of the number by infinite continued fractions; for example:

[1]The reader who would like to know more about these practical applications is referred to *What Is Mathematics?* by Richard Courant and Herbert Robbins [Oxford University Press, 1941].

$$e = 2 + \cfrac{1}{1 + \cfrac{1}{2 + \cfrac{2}{3 + \cfrac{3}{4 + \cfrac{4}{5 + \cfrac{5}{6 + \dots}}}}}}$$

and

$$e = 2 + \cfrac{1}{1 + \cfrac{1}{2 + \cfrac{1}{1 + \cfrac{1}{1 + \cfrac{1}{4 + \cfrac{1}{1 + \dots}}}}}}.$$

The second example, which is called a simple continued fraction because all the numerators are equal to 1, can be expressed even more simply as

$$e = [2, 1, 2, 1, 1, 4, 1, 1, 6, 1, 1, 8, 1, \dots].^\dagger$$

It was also Euler who developed from an earlier discovery the most famous formula in all mathematics, one which expresses the relationship that exists among the three very special numbers—e and i and π—and the two always special natural numbers, 0 and 1:

$$e^{i\pi} + 1 = 0.$$

†The reader who is interested in calculating e from one of the above will find the method detailed in the problem at the end of this chapter.

Reactions to this simple and elegant formula have ranged from that of the American mathematician, Benjamin Peirce (1809–1880), who announced to his Harvard class, "Gentlemen... we cannot understand it, and we don't know what it means, but we have proved it, and therefore, we know it must be the truth," to that of an anonymous mathematician who facetiously defined e as the number that makes the famous equality possible.

By the time of Euler, the invention of analytic geometry by Descartes and the independent invention of the calculus by Newton and Leibnitz had pushed back the traditional frontiers of mathematics and opened up a great new domain. To arithmetic, geometry and algebra now was added *analysis*. In analysis undoubtedly the most important single number is Euler's number, e. This too is appropriate, since Euler was such a master of the new art that he has been described as "Analysis Incarnate."

In analysis we find a quite different approach to the definition of e. It seems particularly complicated for a number when we recall the straightforward "God-given" character of the natural numbers, or even the "man-made" character of such frankly invented numbers as i. For the number e is defined by the area under a certain curve at a certain point—that point being called the number e.

Surely such a number can have nothing to do with 0, 1, 2, 3, ... But, as we shall see very shortly, it has a great deal to do with those numbers.

The necessary curve for the analytic definition of e is determined by graphing all those points for which the vertical or y coordinate is the reciprocal of the horizontal or x coordinate. In other words, all those points (x, y) for which $y = 1/x$. When we mark these points just for the whole number values on the x axis, we see the beginnings of our curve.

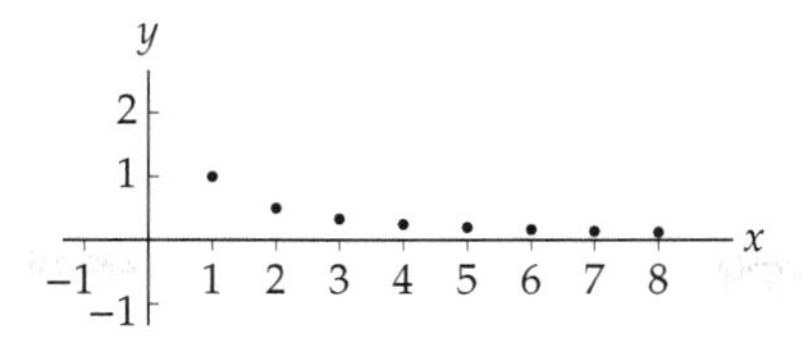

From our illustration it is easy to see that we can make the curve continuous by marking for every real number on the x axis the point on the y axis that is its reciprocal. The number e is then defined in relation to this curve and the area under it between 1 and x.

Suppose we mark off this area by erecting two perpendiculars to the x axis, the left-hand boundary at 1 and the right-hand boundary at some other point x. When this right-hand perpendicular boundary is also erected at $x = 1$, the area under the curve will obviously be 0; but the farther we move x, or our perpendicular boundary, to the right the larger the area under the curve will become.

Here we ask the question upon which the analytic definition of the number e rests. *At what point on the x axis must the perpendicular be erected so that the area under this curve between 1 and x is exactly equal to 1?* We answer this question in the way of mathematicians, not by locating the point, but by naming it. That point along the x axis at which the area becomes exactly 1 is defined as the number e.

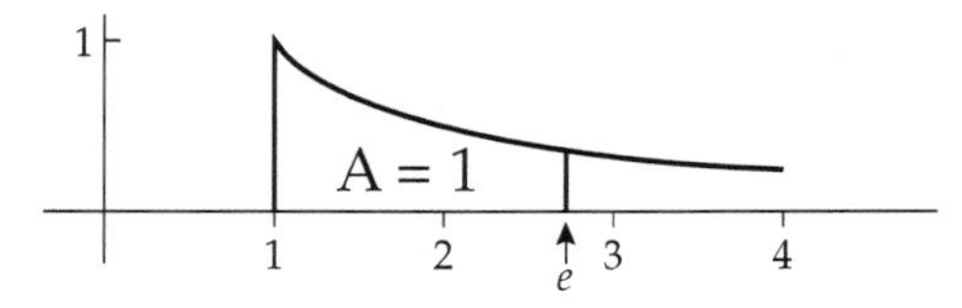

The area under the curve between 1 and x is the logarithm of the real number x to base e.[3]

[3]We can also characterize e in the following down-to-earth manner: if

That there should turn out to be be an intimate relationship between $\log_e x$ and the number of primes below a given number x was one of the most totally unexpected things in all mathematics.

Before we state this relationship, however, let us think back again to the natural numbers with which we began—each one separated from its predecessor by a unit and from its successor by a unit, beginning with zero and continuing to infinity. We recall the first great surprise—that the numbers in this simple regular sequence should fall—in an apparently quite irregular way—into two very different kinds of numbers, the indivisible primes and the divisible composite numbers. The second surprise was that the indivisible numbers should continue, like the numbers themselves, without end; for surely common sense would lead us to expect that when numbers are large enough, they must be divisible by some number smaller than themselves. From the beginning it was always clear that any composite number can be produced by primes, but the third great surprise was that one and only one combination of primes can produce a given number.

These few examples are enough to remind us that the numbers 0, 1, 2, 3, . . . are full of surprises. The relationships we have noted in the past, however, all have one thing in common. Mathematicians have guessed them by examining the natural numbers themselves, and they have proved them by the use of the mathematical properties of the system of natural numbers. No e was allowed here—or needed!

But now we come to a different kind of problem. We have seen that the occurrence of primes in detail is exceedingly irregular. Given a number ending in 1, 3, 7 or 9, we are

---

we invest one dollar at 100 percent interest compounded continuously, at the end of one year we shall have $2.72—that is, e dollars.

not able to tell immediately whether it is prime. Given an already known prime, we cannot tell what the next prime will be. Although we have never been able to prove that there is a point beyond which there are no more pairs of primes separated only by one other number, we know that there are great deserts where we can find in succession as many numbers as we please without a single prime among them. We have also noticed that although the primes continue without end, they become fewer and fewer so that, paradoxically, we can say that although the primes are infinite, almost all numbers are not prime.

Amidst all the irregularity of prime numbers in the small, we find in the large a certain regularity. The decline is slow and steady. There are 25 primes in the first 100 numbers, 168 in the first 1,000, 1,229 in the first 10,000, and 9,592 in the first 100,000. But what is the pattern of this decline?

It was Gauss with his clear eye for patterns among the numbers who first perceived the amazing relationship between the slow, steady decline of the prime numbers and the area between 1 and x under the curve $y = 1/x$ that we illustrated a few pages back. What he conjectured has come to be known as the Prime Number Theorem, the most important truth about the natural numbers discovered in modern times.

To understand the statement of this great theorem (the proof is unfortunately beyond us), we must begin by considering the problem of measuring the density of primes below a given number x. In the first ten numbers we find four primes, 2, 3, 5 and 7. We can represent the density of the primes below 10 by the ratio $4/10$, or .40. In the first 100 numbers we find only 25 primes, the density dropping to $25/100$, or from .40 to .25. If we continue our examination of the numbers we find that after the first 1,000 numbers we have a density ratio of 168/1,000, so that the measure of density has now dropped from .40 to .25 to .168. What Gauss observed was that this ratio approaches closer and closer to $1/\log_e x$.

This relationship can be stated entirely in the symbols

$$\frac{\pi(x)}{x} \sim \frac{1}{\log_e x}$$

where $\pi(x)$ is the number of primes under a given number x so that the ratio $\pi(x)/x$ is the measure of their density, the sign $\sim$ signifies that the two ratios are asymptotically equal, and the ratio $1/\log_e x$ is the reciprocal of the natural logarithm of x to base e.

The relationship so expressed *is* the Prime Number Theorem, usually stated in the following manner:

$$\lim_{x\to\infty} \frac{\pi(x)}{x/\log_e x} = 1.$$

The increasing accuracy of this approximation can best be shown by taking the actual densities for primes under a given x and comparing them to the approximation given by $1/\log_e x$, as in the table below:

$x =$	*Actual Density* $\pi(x)/x$	*Approximate Density* $1/\log_e x$
1,000	0.168	0.145
1,000,000	0.078498	0.072382
1,000,000,000	0.050847478	0.048254942

The extreme difficulty of proving this theorem—it was difficult enough merely to perceive—is shown by the fact that even Gauss, who conjectured it, could not prove it. One of Gauss's last students, G. F. B. Riemann (1826–1866), who although he died at thirty-nine laid the mathematical foundations for the theory of relativity, outlined the strategy of attack on the theorem in a brief, brilliant memoir when he was thirty-three. Riemann could not prove it either.

The Prime Number Theorem differs in a most important respect from the other theorems about primes that we have met earlier in this book. Euclid's great proof that the number of primes is infinite, for example, arises so directly from the natural numbers that anyone with a little calculation among the smallest primes can convince himself of its truth. But no amount of calculation can convince us in the same way of the truth of the Prime Number Theorem. As Hardy once said, "You can never be sure of the facts without the proof." And the proof, unlike Euclid's, comes from outside the natural numbers.

Not until 1896 did mathematicians succeed in proving the Prime Number Theorem. In that year it was proved independently by the French mathematician Jacques S. Hadamard (1865–1963) and the Belgian mathematician C. J. de la Vallée Poussin (1866–1962). This was nearly a century after the relationship that the theorem states was first perceived. Both proofs were of fantastic difficulty, and almost a century of intensive effort has not brought them within the reach of anyone but a professional mathematician.

Since the time of Riemann, the prime number theorem has been the central problem of what is known as the analytic theory of numbers, a discipline that uses the most advanced methods of the calculus and is considered from a technical point of view one of the most difficult in all mathematics. Unlike the classical number theory of the Greeks, it does not limit itself to the natural numbers in its efforts to uncover the relationships and interrelationships that exist among 0, 1, 2, 3,.... It brings to the battle infinities upon infinities of other numbers: rationals and irrationals, positive numbers and negative numbers, real numbers and imaginary numbers. All are marshalled together under the banner of *the complex numbers*. These are numbers of the form $x + yi$, half-real and half-imaginary. When $x = 0$, they are pure imaginaries. When $y = 0$, they are real numbers.

Against this formidable array, the natural numbers would seem to be both literally and figuratively outnumbered; but although they yield much, they also retain much. What we find out about them through the use of these other numbers is found only with great difficulty. What we win is never easily won.

In the story of the relationship between a number like e, the base of the natural logarithm, and the numbers 0, 1, 2, 3,..., we catch a glimpse of the underlying unity of all numbers. Like the relationships among the natural numbers that are inherent in the simple sequence 0, 1, 2, 3,..., the relationships among the different kinds of numbers are inherent in the structure of that imposing edifice, the concept of number, which has been erected, block by block, upon the foundation of the natural numbers themselves.

The wonder is not merely that the relationships are there, but that they are so very difficult for us to perceive—and to prove.

· ANOTHER PROBLEM ·

The infinite continued fraction by which Euler represented e is defined as the limit of the sequence

$$2, 2+\frac{1}{1}, 2+\cfrac{1}{1+\cfrac{1}{2}}, 2+\cfrac{1}{1+\cfrac{1}{2+\cfrac{2}{3}}}, 2+\cfrac{1}{1+\cfrac{1}{2+\cfrac{2}{3+\cfrac{3}{4}}}}, 2+\cfrac{1}{1+\cfrac{1}{2+\cfrac{2}{3+\cfrac{3}{4+\cfrac{4}{5}}}}}, \ldots.$$

The reader may enjoy computing these first few terms and seeing how the sums, alternately smaller and larger, approach closer and closer to the value of e.

$$e > 2$$

$$e < 2+\frac{1}{1} = 3$$

$$e > 2+\cfrac{1}{1+\cfrac{1}{2}} = 2+\frac{1}{3/2} = 2\frac{2}{3} \text{ or } 2.67$$

$$e < 2+\cfrac{1}{1+\cfrac{1}{2+\cfrac{2}{3}}} = 2+\cfrac{1}{1+\cfrac{1}{8/3}} = 2+\frac{1}{1+3/8} = 2\frac{8}{11} \text{ or } 2.73$$

$$e > 2+\cfrac{1}{1+\cfrac{1}{2+\cfrac{2}{3+\cfrac{3}{4}}}} =$$

$$e < 2+\cfrac{1}{1+\cfrac{1}{2+\cfrac{2}{3+\cfrac{3}{4+\cfrac{4}{5}}}}} =$$

· ANSWER ·

$e > 2\frac{38}{53}$ or 2.717, $e < 2\frac{74}{103}$ or 2.7184.

· ALEPH-ZERO ·

Zero, with which we began our story of the numbers, was the most practical invention in the history of mathematics. The theory of infinite sets, which we are now going to take up, may well be the most impractical; yet from the point of view of mathematics, it is incomparably the more important.

Although the modern mathematical theory of the infinite is not properly a part of the theory of numbers, it permeates the modern theory (as it does all of modern mathematics) and develops quite naturally from a consideration of the numbers with which we have been concerned in this book. We have seen in the preceding chapters that the mathematically interesting sequences of numbers are those that continue without end. If the primes were finite, they would be of considerably less interest; and if it is established ultimately that the perfect numbers are finite, their interest will become merely historical. Odd and even numbers, the primes and the composite numbers, the squares, the cubes, the curious pentagonal numbers, all are infinite. These infinite sequences of numbers among the infinite sequence of the natural numbers first suggested the revolutionary idea that is the cornerstone of the modern theory of the infinite.

To understand this idea, we have only to go back to Galileo, who held the cornerstone in his hands but failed to put it into place. In "Four" we told how he, in the character of Salviatus, pointed out that there are fully "as many" in the infinite set of squares as there are in the infinite set of all numbers. His argument was simplicity itself. Every number, by definition, has a square, which is that number multiplied by itself. We can pair the first square with the first number, the second square with the second number, and so on. We shall never run out of squares until we run out of numbers; and since we shall never run out of numbers, we shall never run out of squares. In a similar way we pair the fingers of the right hand with those of the left, right thumb to left thumb, right forefinger to left forefinger, and so on; when we come out even, we say we have "as many" fingers on one hand as we have on the other.

Galileo did no more than to extend this commonly accepted way of determining "as many" to infinite quantities. He pointed out that there is a square for every number throughout the entire sequence of numbers. Squares and numbers can be paired "to infinity." In spite of appearances to the contrary, there are as many squares as there are numbers.

When we say that Galileo did no more than extend the commonly accepted way of determining "as many" from finite to infinite quantities, we do not intend to minimize his achievement; for in some two thousand years no other mathematician did so much. But Galileo, having come so close to the modern theory of the infinite, did no more.

He had showed that logically there are as many squares as there are numbers; then he asked himself the next question. *If there are as many squares as there are numbers, can the number of squares be said to be equal to the number of numbers?*

Well, how was a mathematician going to answer that one?

If there are as many squares as numbers, as he himself had shown, the two sets cannot be said to be unequal. On the other hand, there are obviously many more numbers that are not squares than there are numbers that are squares. At this point Galileo put the cornerstone back on the rock pile and concluded, as we have seen already in "Four":

"I see no other decision that it may admit, but to say that all Numbers are infinite; Squares are infinite; and that neither is the multitude of Squares less than all Numbers, nor this greater than that; and in conclusion, that the Attributes of Equality, Majority, and Minority have no place in Infinities, but only in terminate quantities."

Three hundred years later, the mathematician Georg Cantor (1848–1918) recognized that inherent in the definition of "as many" were the concepts of "equality" and of "the same number." To apply to infinities these concepts, usually applied only to finite quantities, he needed a truly precise definition of an infinite set as opposed to a finite set. Such a definition he found in the relationship that Galileo had earlier perceived between the squares and all the numbers although Cantor did not in fact come to his discovery from the same direction as Galileo.

An infinite set, so Cantor defined it, *is one that can be placed in one-to-one correspondence with a proper part of itself.*

This definition obviously does not apply to a finite set. Although we can place all the squares in one-to-one correspondence with *all* the numbers, never running out of either squares or numbers, we cannot place the squares less than ten in one-to-one correspondence with the numbers less than ten for the simple reason that we run out of squares before we run out of numbers.

COUNTING ALL SQUARES			COUNTING THOSE UNDER TEN	
0	0		0	0
1	1		1	1
2	4		2	4
3	9		3	9
4	16	but	4	
5	25		5	
6	36		6	
7	49		7	
8	64		8	
9	81		9	
...	...			

At this point we may well ask ourselves why, if Galileo perceived the essential characteristic of an infinite set as distinguished from a finite set, did he not go on to Cantor's theory of the infinite, three hundred years before Cantor? The answer to this question lies in one of the most ancient axioms of mathematics: the axiom, found in Euclid's *Elements*, to the effect that the whole is greater than the part. Galileo could not bring himself to deny this axiom by saying that the whole (all the numbers) is equal to the part (all the squares). Instead he decided that the Attribute of Equality had no place in infinite quantities. Cantor said in essence that the axiom that the whole is greater than the part *has no place in infinite quantities.*

The mathematical justification for Cantor's revolutionary reversal of Euclid's axiom lies, very simply, in the fact that the reversal *works* with infinite quantities; that is, it does not lead us into contradictions. On the other hand, the axiom that the whole is greater than the part, which works for finite quantities, leads us into contradictions when we apply it to infinite quantities. Before Cantor, mathematicians had struggled in vain to resolve these contradictions. Cantor, defining an infinite set as one that, unlike a finite set, can be placed in one-to-one correspondence with a part of itself, resolved them all by eliminating them.

His theory of the infinite is famous for many *seeming contradictions*. We can prove, for instance, that there are as many points on a line one inch long as there are on a line one mile long; we can also prove that there are in all time as many years as there are days.[1] But we never find

[1]To prove that there are as many points on the short line as there are on the long line, we take the line *AB* and the longer line *CD*, place them parallel to each other, and join the ends *AC* and *BD*. We extend *AC* and *BD* until they intersect at *O*. It is then easy to see that any line drawn from *O* through the two lines *AB* and *CD* will intersect them at the points *P* and *Q*, respectively. For every point *Q* on the longer line there will be a point *P* on the shorter that can be paired in one-to-one correspondence with it.

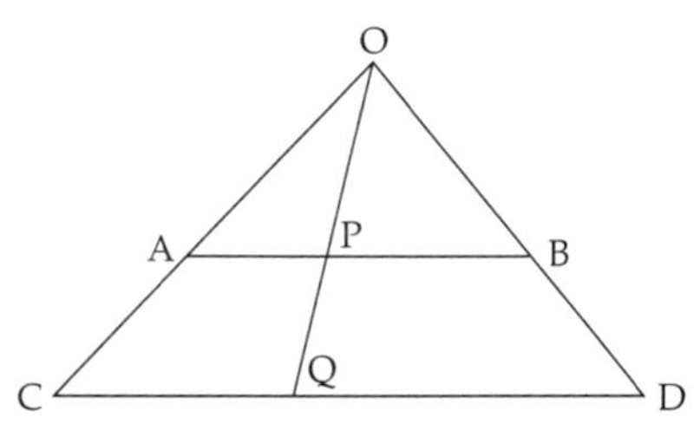

That there are in all time as many years as there are days is what Bertrand Russell calls the Tristram Shandy paradox. Shandy, we recall, spent two years recounting the events of the first two days of his life and bemoaned the fact that at this rate he would fall farther and farther behind in his autobiography. Quite true for a mortal Shandy. But an immortal Shandy, with all eternity at his disposal, would recount the first day's events in the first year, the second day's in the second year, and so on; and eventually he would arrive in his narrative at any given day.

ourselves in the untenable position of having proved in both cases mentioned that the compared sets are equal—and that they are unequal. The justification of consistency—that an axiom does not lead to self-contradictory statements—is all the justification a mathematician needs. By the rules of the game, he is then free to formulate any theory that follows logically from his axiom.

This is exactly what Georg Cantor did. Having defined an infinite set as one that can be placed in one-to-one correspondence with a part of itself, he then defined infinite sets that can be so paired as *equal* and as *having the same number*.

Any infinite set that can be placed in one-to-one correspondence with the set of all the positive integers has *the same cardinal number* as the set of all the positive integers. It is not the last positive integer, for there is no last. It is the number of the totality of positive integers. It is a cardinal number because it answers the question *How many*? about the set of positive integers just as two and three answer the question *How many*? about the sets of pairs and triplets to which they apply. But it is an entirely new kind of cardinal because it answers the question, not about finite, but about infinite sets. Cantor called it a transfinite cardinal and boldly presented it with a name. As the Greeks had called their numbers by the letters in their alphabet, he called his after the first letter of the Hebrew alphabet, *aleph*, the symbol for which is $\aleph$.

Up until Cantor's time infinity, represented by three dots at the end of a sequence of numbers or by the symbol ∞, had been the ultimate in unfinished business: an ever-increasing finite quantity—add one and you always got a larger quantity, a larger number—there was no last number. Of course, Cantor actually changed none of this. The *aleph* of the positive integers is no last number. It is, for one thing, not a positive integer at all. The relation of the transfinite cardinal of the positive integers to the integers themselves is somewhat similar to the relation of the number one to the proper fractions. One is not itself a fraction; it is the limit that the fractions approach. No matter how large a fraction we choose (that is, how close to one in value), there is always another fraction that is larger than the first and, therefore, closer to one; yet there is no such fraction that exceeds or equals one. In very much the same way this particular *aleph*, which is not a positive integer itself, is the limit of all the positive integers. No matter how large the integer we choose, there is always another that is larger, although it is not any "closer" to the limit. The essential difference, for this example, between the number one and the number *aleph* of the positive integers is that while the fractions literally approach the limit one, the positive integers approach their transfinite cardinal only because the larger they get the farther they get from zero. No matter how large the integer we choose, we never get any "closer" to infinity because between us and infinity is always an infinity of numbers equal to the infinity of positive integers.

The idea of infinity, not as something that is in the process of becoming, but as something that exists—a number that can be handled in many ways just like a finite number: added, multiplied, raised to a power—was as revolutionary as any idea that has ever flowered in the mind of man. Like all revolutionary ideas, it was opposed with emotion, much of it blind and bitter. Even Gauss, who thought far ahead of his time and disembarked on many mathematical shores long before their official discoverers, could not accept the idea of a *consummated* infinite.

Probably no great mathematician ever stood more completely alone with his idea than Cantor, but he stood firm:

"I was logically forced, almost against my will, because in opposition to traditions which had become valued by me, in the course of scientific researches, extending over many years, to the thought of considering the infinitely great, not merely in the form of the unlimitedly increasing... but also to fix it mathematically by numbers in the definite form of a 'completed infinite.' I do not believe, then, that any reasons can be urged against it which I am unable to combat."

Cantor's confidence lay not only in his mathematics but in mathematics itself. He was always aware of the inherent freedom of mathematical thought, and at another time he wrote:

"... mathematics is, in its development, quite free, and only subject to the self-evident condition that its conceptions are both free from contradiction in themselves and stand in fixed relations, arranged by definitions, to previously formed and tested conceptions. In particular, in the introduction of new numbers, it is only obligatory to give such definitions of them as will afford them such a definiteness and, under certain circumstances, such a relation to the older numbers, as permits them to be distinguished from one another in given cases. As soon as a number satisfies all these conditions, it can and must be considered as existent and real in mathematics."

Cantor did not fear such freedom. He recognized that the conditions laid down for it were very strict, arbitrary abuse being kept at a minimum. He recognized also that unless a new mathematical conception was mathematically useful, it was abandoned in short order. Both the mathematical soundness of Cantor's conception of a consummated infinite and the mathematical usefulness of his transfinite cardinals have been borne out by time. Even before his death in 1918, his ideas had been quite generally accepted; and the arithmetic of the transfinite cardinals that we shall detail briefly in the next few pages is now as much a part of mathematics as 2×2.

Having defined cardinality for infinite quantities, Cantor proceeded to make two important statements: (1) the cardinal number of all the positive integers is the smallest transfinite cardinal and (2) for every transfinite cardinal there exists a *next* larger transfinite cardinal. The similarity between the totality of transfinite cardinals and the totality of everyday finite cardinals, or the natural numbers, is apparent. There is a first; there is always a next; there is no last.

All of these transfinite cardinals were called *alephs* by Cantor, but to each he added a subscript that indicated its place in the sequence. The number of the positive integers, first and smallest of the transfinite cardinals, has zero as its subscript and is signified by $\aleph_0$; in words, it is "*aleph*-zero." The next largest transfinite cardinal is *aleph*-one; the next largest *aleph*-two, and so on.[2] The reader should not, however, conclude that this unlimited sequence of transfinite cardinals exhausts the transfinite cardinal numbers, for there exists a number that is the sum of all these *aleph*s and "out of it," as Cantor wrote, "proceeds in the same way... a next greater..., and so on, without end."

Is there an everyday example of an infinity that is larger than that represented by *aleph*-zero; in other words, an infinity that cannot be placed in one-to-one correspondence with the positive integers?

It was Cantor's achievement, not only to produce such an infinity, but to produce it by a method so simple that a person with no more knowledge of the theory of the infinite

[2]Interestingly, although Cantor later came to *aleph*-zero, he initially began to number his *aleph*s with one rather than zero.

than that we have been able to expound in the few pages of this chapter will have no trouble in following his proof. To appreciate his achievement, however, we must realize that just as the positive integers can be placed in one-to-one correspondence with one of their subsets, such as the squares or the primes, infinities that include the positive integers themselves as one of their subsets can also be placed in one-to-one correspondence with them.

An example of such an apparently larger infinity that actually has the same cardinal number as the positive integers is the infinity of all positive rational numbers. The rational numbers, as we recall from "One," include all those quantities that can be represented by the ratio of two whole numbers. Since the quantity represented by the ratio of a whole number to one is that number itself as $2/1 = 2$, the rationals include the positive integers as well as well as what we call fractions. Intuition tells us that there are many more rationals than integers, but intuition also tells us that there are fewer squares than there are integers. And intuition is not mathematical proof. If we can count the rational numbers, we can pair a given fraction a/a_l with 1, a second b/b_1, with 2... so that in a finite amount of time we shall be able, if we wish, to count to any given rational number.

Well, let us begin. But how?

There is no smallest rational number.

There is no next largest rational number.

Given a rational number a/b as "the smallest," we can always get a smaller by adding one to the denominator, $a/(b+1)$ being smaller than a/b. Given any two rational numbers, a/b and c/d, no matter how close together, we can always produce another that lies between them by adding the numerators and the denominators of both for a new rational number. Between a/b and c/d lies $(a + c)/(b + d)$.

Have we then found an infinity that is larger than the infinity of positive integers, represented by *aleph*-zero? No, we have not. For it is possible to arrange the rational numbers (although not according to increasing size) in such a way that there is a unique rational number to be paired with each of the positive integers and that, given a sufficient but finite amount of time, we can count to any rational we choose.

We begin by arranging all the rational numbers in subsets according to their numerators, omitting all those with common factors since they will already have been included. We now have an infinite number of rows of rational numbers, and it is obvious that the rows can be placed in one-to-one correspondence with the positive integers:

$$\begin{array}{cccccccc}
1 & \leftrightarrow & 1/1 & 1/2 & 1/3 & 1/4 & 1/5 & \ldots \\
2 & \leftrightarrow & 2/1 & 2/3 & 2/5 & 2/7 & 2/9 & \ldots \\
3 & \leftrightarrow & 3/1 & 3/2 & 3/4 & 3/5 & 3/7 & \ldots \\
\vdots & & \vdots & \vdots & \vdots & \vdots & \vdots &
\end{array}$$

But each column also contains an infinite number of rational numbers that can also be placed in one-to-one correspondence with the positive integers:

$$\begin{array}{ccccccc}
1 & 2 & 3 & 4 & 5 & 6 & \ldots \\
\updownarrow & \updownarrow & \updownarrow & \updownarrow & \updownarrow & \updownarrow & \\
1/1 & 1/2 & 1/3 & 1/4 & 1/5 & 1/6 & \ldots \\
2/1 & 2/3 & 2/5 & 2/7 & 2/9 & 2/11 & \ldots \\
3/1 & 3/2 & 3/4 & 3/5 & 3/7 & 3/8 & \ldots \\
\vdots & \vdots & \vdots & \vdots & \vdots & \vdots &
\end{array}$$

We have here infinities upon infinities. If we count by rows, we shall never get to the end of the first row and, therefore, never to the beginning of the second row, or the rational number 2/1. If we count by columns, we shall never get to the end of the first column and, therefore, never to the beginning of the second column, or 1/2.

Yet there is a way of counting the rationals so that we can place each of them in one-to-one correspondence with a unique positive integer and so that, in a finite time, we can count to any given rational number. We can do this—Cantor showed—by counting the same arrangement of rows and columns *on the diagonal*:

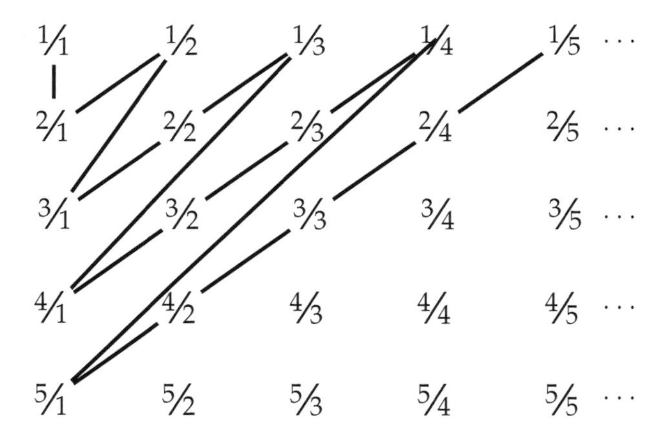

By this method we have a first rational number to be counted (1/1) and we always have a next number (in the illustration above, after 1/5 we go next to 6/1). We have no trouble at all in getting to 2/1 and 1/2. It is apparent that given sufficient time we can count to any rational number we choose. We shall never run out of numbers with which to count. There are, in spite of appearances to the contrary, fully as many positive integers as there are rational numbers.

1	$\frac{1}{1}$
2	$\frac{2}{1}$
3	$\frac{1}{2}$
4	$\frac{3}{1}$
5	$\frac{2}{3}$

and so on.

The transfinite cardinal number of both sets is the same. It is *aleph*-zero. Such sets, which can be placed in one-to-one correspondence with the positive integers (i.e., can be "counted" by them), are said to be "denumerable."

After a result so contrary to intuition, can we, in accordance with Cantor's claim that for every transfinite cardinal there is a next transfinite cardinal, now place on exhibition a set that is "larger"—in short, a set that is "non-denumerable," that cannot be "counted" by the positive integers? Indeed we can. Cantor himself showed that the infinity of decimal fractions between zero and one cannot be placed in one-to-one correspondence with the positive integers and must, therefore, have a cardinal number greater than *aleph*-zero.

The decimal fractions include both the rational numbers and the irrational numbers, such as the square root of 2, that cannot be represented by the ratio of two whole numbers. Among the decimal fractions there are some that terminate in a string of zeros, others that after a certain sequence of numbers begin to repeat that sequence indefinitely, and still others—those that represent the irrational quantities—that by their nature never terminate in zeros and never repeat. All three types can be considered as non-terminating (the string of zeros in the first type continuing to infinity), and all can be represented by the general form

$$0.n_1n_2n_3n_4n_5n_6n_7n_8n_9\ldots$$

where each n represents a given place in the decimal.

Just as it is impossible to write down the first rational number greater than zero, it is also impossible to write down the first decimal fraction; and just as it is impossible to write down the next rational number, it is also impossible to write down the next decimal fraction. Yet it was possible to arrange the rational numbers in such a way that there was a first to be counted, and a next, and so on; and we could arrive in a finite length of time at any given rational number. Is there a similar way to arrange the decimal fractions so that they too can be placed in one-to-one correspondence with the integers?

Mathematics allows two ways of answering this question: produce an arrangement, or show that no such arrangement is possible. Cantor did the latter, and did it as simply and smoothly as Euclid two thousand years before him had proved that the number of primes is infinite. We recall (from "Three") that Euclid began by assuming a finite set that included all the primes; he then showed that by multiplying the primes of the set together and adding one, he could always produce either a prime not included in the set or a number the prime factors of which had not been included. The assumption then that there could be a finite set of all primes was shown to be false; the primes, infinite.

This is exactly the method Cantor followed. To prove that the set of all decimal fractions between zero and one cannot be placed in one-to-one correspondence with the positive integers—and is thus non-denumerable—he assumed that by some unspecified arrangement such a correspondence was possible. He assumed a first decimal fraction determined by this arrangement and paired it with the first positive integer. He then assumed a next and paired it with the second positive integer, and so on.

$$\begin{array}{ccl} 1 & \leftrightarrow & 0.a_1a_2a_3a_4a_5a_6a_7a_8a_9\ldots \\ 2 & \leftrightarrow & 0.b_1b_2b_3b_4b_5b_6b_7b_8b_9\ldots \\ 3 & \leftrightarrow & 0.c_1c_2c_3c_4c_5c_6c_7c_8c_9\ldots \\ \ldots & & \ldots \end{array}$$

He then showed that such an assumption of one-to-one correspondence between the decimals and the positive integers is false, because he could always produce a decimal fraction that had not been counted. The uncounted decimal fraction he represented as

$$0.m_1m_2m_3m_4m_5m_6m_7m_8m_9\ldots$$

where m_1 is a digit other than a_1 in the "first" decimal; m_2, a digit other than b_2 in the "second" decimal; m_3, a digit other than c_3 in the "third" decimal; and so on.[3] A decimal fraction formed in this manner could not be included in the assumed arrangement of "all" decimal fractions because it differs from each fraction that has been included in at least one place. Thus in this way the infinity of decimal fractions is shown to be greater than the infinity of positive integers: the two sets cannot be placed in one-to-one correspondence. And since it is greater, its cardinal number must be greater.

The decimal fractions between zero and one cover only an infinitesimal part of the number line—the continuum of

[3]Since terminating decimals like 0.25 can be represented as nonterminating decimals in two ways: either as 0.25000... or as 0.24999..., Cantor excluded the digit nine to avoid having the new decimal a different representation of a number that had already in a different form been included in the class of "all" decimals.

real numbers that provides a number for every point on the line. But what is true of that part of the line is true of the continuum as a whole. Cantor, therefore, called his new transfinite cardinal "the number of the continuum" and took as its symbol a letter from the German alphabet.

This departure from the Hebrew alphabet was significant, for Cantor could not establish where among the *aleph*s the number of the continuum stands. He had stated that *aleph*-zero was the smallest transfinite cardinal, that for every transfinite cardinal there was a next largest and, most important, that *all* the transfinite cardinals were included in the sequence of *aleph*s. Now he had produced in the continuum of real numbers a set of numbers with a transfinite cardinal other than *aleph*-zero and larger than *aleph*-zero, but was it the *next aleph*—was it *aleph*-one? This was the question that Cantor left to the mathematicians that followed him. It had all the beguiling simplicity of one of the Greeks' questions about the natural numbers, but when it was answered—as it finally was—the answer was one that many mathematicians were to find highly unsatisfactory.

Cantor himself always believed that the number of the continuum was indeed *aleph*-one. Proving this conjecture of his—the so-called "continuum hypothesis"—was to become one of the great mathematical challenges of the twentieth century.

Unfortunately Cantor himself had not *proved* all the statements he had made about the *alephs*, although they would follow logically *if* (as Cantor believed) all infinite sets could be "well-ordered"—that is, ordered in such a way that every nonempty subset had a first element. But this theorem was proved only later by Ernst Zermelo (1871–1953), and to prove it he had to invoke a new axiom—what is known as *the axiom of choice*.

In very nontechnical language the axiom of choice assumes that it is possible to make infinitely many choices even when one has no rule for choosing. A popular example among logicians is the case of the man with a denumerable infinity of pairs of shoes and pairs of socks that he wants to "count" by placing in one-to-one correspondence with the positive integers. To do so, he must begin with a "first" shoe and a "first" sock from each pair. He has no problem deciding which will be the first shoe in a pair, since there is a right and a left, but by what rule can he choose a first sock from each pair? According to the axiom of choice, he can take whichever sock he wishes—that *is* the rule. Those mathematicians who do not accept the axiom of choice say that he can't assume he can choose unless he has a rule for choosing. Those—the majority—who favor the axiom say he can choose, rule or no rule.

Since the time of Cantor the theory of infinite sets has been rigorously axiomatized after the model established by Euclid in his *Elements*. Theorems in set theory, like the theorems of geometry, have to be logically derived from a small group of accepted assumptions, or *axioms*, and from theorems that have been derived previously from those same axioms. There has, however, always been some doubt whether the axiom of choice should be included among the axioms. Without it many important and beautiful theorems about infinite sets cannot be proved; yet it makes a statement about infinite sets that, in mathematical language, is "not constructive." Such a statement makes many mathematicians uncomfortable. To put it bluntly, even those who utilize the axiom of choice in their proofs would be happier if they didn't have to do so. Unfortunately the axiom of choice is necessary even to prove Cantor's statement that the cardinal number of the continuum is in fact *one of the alephs*—a statement that is obviously much weaker than his hypothesis that the cardinal number of the continuum is *aleph*-one.

Mathematicians who have come after Cantor have always believed with him that the number of the continuum was in fact *aleph*-one, the next *aleph* after *aleph*-zero. They have also believed that ultimately one of them would prove, from the axioms of set theory, that it was—or in the alternative but also from the axioms, that it was not.

"Take any definite unsolved problem[s]..." the great German mathematician David Hilbert (1862–1943) said in a famous talk at the beginning of the twentieth century, "however unapproachable they may seem to us and however helpless we stand before them, we have, nevertheless, the firm conviction that their solution must follow by a finite number of purely logical processes."

Hilbert's conviction "that every definite mathematical problem must necessarily be susceptible of an exact settlement, either in the form of an actual answer to the question asked, or by the proof [as in the case of such problems as the trisection of the angle] of the impossibility of its solution and therewith the necessary failure of all attempts" was to be severely tested in the coming century—and in connection with the questions posed by the axiom of choice and by Cantor's continuum hypothesis.

In 1938 Kurt Gödel (1906–1978) showed that the axiom of choice *cannot be disproved* from the other axioms. Twenty-five years later the young American mathematician, Paul Cohen (1934–), showed that it *cannot be proved* from the other axioms. This result established that there is no way around the axiom of choice. It is *consistent* with the other axioms (Gödel) and *independent* of them (Cohen). If you need it, you must include it as an axiom.

More unsettling to mathematicians was another result of Cohen's. Gödel had showed that the continuum hypotheses cannot be disproved from the axioms of set theory. Cohen now showed that it cannot be proved from the axioms. In the language to which mathematicians have come since the time of the Greeks, and indeed since the time of Hilbert, the question posed by the continuum hypothesis is *undecidable.*

Many mathematicians, including Gödel himself, found such an answer highly unsatisfactory. If mathematics is, as one mathematician has called it, "the ultimate reality," then either the number of the continuum is *aleph*-one or it is not *aleph*-one. Some mathematicians referred to Cohen's solving of the problem of the continuum hypothesis as its "unsolving." But Cohen himself feels that it is the best answer that can be obtained.

Two thousand years ago, as we saw in "Six," mathematicians asked, *How many numbers are there that are the sum of all their divisors?* Are they finite or infinite? Today they would like to know with the traditional certainty of mathematics whether the number of the continuum is *aleph*-one. Both questions remain unanswered, but between the askings lie two thousand years under the spell of an apparently simple sequence that begins with zero and continues without end.

· THE ARITHMETIC OF THE · · INFINITE ·

The arithmetic of transfinite numbers is as paradoxical as the theory of the infinite itself. Questions in the "two plus two" class of ordinary arithmetic are either so simple that they are trivial or so difficult that no one has been able to answer them.

The reader will find that he can answer the problems below if he will recall what he has learned about the various infinite sets among the real numbers, and he can "show" that he is right by producing an example from the text:

$$\aleph_0 + \aleph_0 =$$
$$2 \times \aleph_0 =$$
$$\aleph_0 \times \aleph_0 =$$

· ANSWERS ·

The answer to all three problems is the same—*aleph*-zero. Examples could be (1) the odd and the even numbers, (2) the positive and the negative integers, (3) the rational numbers.

· INDEX ·